AN AUTOBIOGRAPHY

1920–2000

Alvin F. Meyer, Jr.

ATHENA PRESS
LONDON

AN AUTOBIOGRAPHY: *1920–2000*
Copyright © Alvin F. Meyer, Jr. 2002

All Rights Reserved

ISBN 1 930493 75 4

First Published 2002 by
ATHENA PRESS
Queen's House, 2 Holly Road
Twickenham TW1 4EG

Printed for Athena Press

AN AUTOBIOGRAPHY
1920–2000

Contents

Chapter I   11

Chapter II   14

Chapter III   17

Chapter IV   21

Chapter V   28

Chapter VI   32
   OFFUT Air Force Base   32
   At SAC   32
   Maintenance Personnel   34
   Industrial College   35

Chapter VII   49
   Asbestos Removal from the Dome of the Library of the Navy Astronomical Center at the Naval Observatory   51
   Visibility Study in the New Design for the White Oak Paint Shop   51
   Correction of Occupational Safety and Health Deficiencies Navy-Wide   52
   The Design of the Navy's Hazardous Substitution Manual   52
   The Effects of Noise on Whales   53
   Navy Manual for the Navy Requirements of Hazardous Materials Control and Management   53

Coast Guard's Hazardous Materials Usage 53

Engineering Studies for the Air Force 54

Air Force Regulations
and Procedures on System Safety 55

Analysis of a Proposed Twin-Screw Extruder
and Explosive 56

Analysis of the Improvements to the OTTO Fuel 56

Last-Cycle Support of the Navy's Deep Water
New Attack Submarine Program 56

Work on the Contract of the Fossil Energy Program
of the Department of Energy 57

Chapter VIII 58

Chapter IX 62

Chapter X   Miscellaneous 64

An Unexpected Find 64

Random Thoughts 65

Epilogue 76

List of Appendices

Appendix I   Deliverable Work Done Under
              NAVSUP Supply System   82

Appendix II   Other Works Done   88

Appendix III An Independent Analysis and
              Recommendations – Implications for
              the DOD on Future US Oil
              Reserve Shortages   93

  Note   93
  Abstract   93
  Introduction   95
  Scope of Impacts on Forces   95
  Implications of Possible Scenarios   96
  The Deterrent Option   97
  Conclusion and Recommendations   98

Addendum   100
  Update: Kosovo Target System   100
  The Importance of Synthetic Fuels   100

Appendix IV   Examples of Projects   102

Appendix V   Examples of Excerpts of Previous
              Recommendations   104

  Pre-1900   104
  1990-1994   105
  1994-2000   105

List of Figures

Figure 1	Worldwide Distribution of Oil	94
Figure 2	Flow Chart of the Hazardous Material Substitution Process	107
Figure 3	A. F. Meyer & Associates, Inc. Organization Chart	108

Chapter I

This chapter is based on a pictorial record and not my personal observations. It is gleaned from a number of sources including photographs taken by my mother. From the time I was an infant in the arms of my mother and father until I was about five years old, I have no recollection of things. However, pictures taken by my mother and my Uncle Lionel are adequate to portray my childhood.

I was born of Alvin Meyer and Bertha Weil-Meyer of Shreveport, Louisiana, in 1920. When I was born, there was a definite air of growth in the South. My father was wounded in WWI after having served as a sergeant in a field hospital supporting the American First Division. He was gassed at Chateau-Thierry. My mother Bertha was deeply in love with her husband, as is seen in these pictures. I have no recollection at all of my father. He died when I was five years old.

My father was the youngest son of his father Abe Meyer. The first child Lionel was later to have a profound influence on my early years. My mother Bertha was the second child of four belonging to her father and mother.

Shortly after my mother and father were married, the effects of WWI gassing became apparent. The next two years were spent in many different places. I went along with them on their many travels, including Michigan.

My father, having graduated from Tulane University, was a sergeant in the Army Field Hospital.

My mother decided to go and live with her parents after the death of my father. This was a normal occurrence for a young widow at that time. My grandfather, Abe, on my father's side, lived in Hamilton Terrace, not far from my grandfather on my mother's side. They lived at 519 Fannin Street. I had a happy time growing up because my Uncles Milton and Albert Henry were still living at home along with my Aunt Minnie. It was a good family in a very fine location.

Several things happened while I was living with my grandparents, which made a profound impression on me. In 1926 when I was six years old, we attended Virginia Military Institute (VMI) for my Uncle Albert Henry's graduation. Then and there, I made up my mind that I would attend VMI. We attended the sesquicentennial in Philadelphia immediately following the graduation. At that event I saw the NC4, which was a flying boat. It made a lasting impression on me.

In 1927, the Red River flooded. During the flood, my Uncle Albert Henry got lost for a day. My maternal grandfather, Harry Weil, was a distinguished citizen and chairman of the Louisiana Red Cross. At the back of his store, I remember, there was storage of tents and mess gear and other equipment that was taken to the fairgrounds and made into a tent city. In 1929, my grandfather Abe died.

Abe was a millionaire having come from France just after the Franco-Prussian War with absolutely nothing. From nothing, he built up a tremendous fortune with his brother Marx Meyer.

Abe left his estate to his daughter Beulah, his son Lionel and myeself, Alvin Jr. My mother Bertha was made my guardian until I was declared an adult. Tom Stagg Sr., Beulah's husband, was selected by the trustees of my grandfather, Abe Meyer's estate to be the managing director, all of which made very little impression upon me as a child. I simply went about my business of playing with my friends, including Molly Ellerbe, who lived across the street. Abe's death, however, marked a change in all of our fortunes.

The Depression started in October of 1929. Abe had died earlier that year in June. The consequences of his death were soon to be seen.

I was not allowed to attend the family gatherings on Sundays until I was nine years old. These family gatherings were very large and included all of the family from the Florsheims and the Weils. Sunday dinners were a big event. There was a large dining room table that extended out into the other room. On Sundays, we had a large spread of food and family conversation. My grandfather, Harry, had built the Cross Lake. He had located on it many fishing spots, and on Sundays, he would go there with my Uncle Milton, who lived with us on Fannin Street with his wife Helen for several years while his new house in Broadmoor Subdivision was being built.

Chapter II

From the time I was six years old until about nine, I attended several schools. I was a graduate of Ms. Jacks School, which was a private school. I also attended the Creswell Street School. I had to take a long streetcar ride to get to school during 1929 to 1933. I have little memory of what went on in school. However, I had an old friend Grace Ingersol, the daughter of Captain and Mrs. Ingersol. Captain Ingersol was a steamboat captain. I also made friends with a number of people, whom I continued to associate with throughout my grade and high-school time. These included Jeanette Sentel, Reese Jones and Calhoun Allen. You will meet them again later in this book when we will discuss them in the right context.

The Depression did not have much of a consequence in Shreveport initially. The city was the second largest in the state and was growing in spite of the hard times associated with the Depression. I well remember the people passing by, who stopped to take the handout that my grandmother and mother always had for them.

I left Creswell Street School to attend C. E. Byrd High School. I entered the band in my freshman year. I became a prominent leader of the band in my sophomore year and continued throughout. Dr. Grover C. Kaufman was the principal. He was a close personal friend of my grandfather, Harry. Dr. Kaufman made a magnificent contribution to my background. I became a debater, leader and I was elected to the Honor Society in my junior year.

One of the things which was my near undoing, was the Junior ROTC. Aaron Selber, president of Selber Brothers, and his brother, both veterans of foreign wars and the American Legion, helped me get into ROTC in Byrd High. I, being the son of a WWI hero, wanted to be active as well. I wanted to participate in student formations. Those in the student body and the families were opposed to a Junior ROTC program. Had it not been for Dr. Kaufman, I probably would have been expelled. However, we prevailed by keeping on with our program until we got ROTC into the school.

In my senior year, I was Cadet Captain of Company C of the ROTC program. I made Grace Ingersol my company sponsor. Jeanette Sentel and Ama Norflet were other friends who were sponsors of other battalion companies.

It was during this time that I was the stage manager of the C. E. Byrd High School Drama Club. I had to do numerous civic events and I was present at all of these.

During this period, I also became interested in the newspaper. John Ewing owned the *Shreveport Times* and was a close friend of my mother. My home on Fannin Street was not far from the newspaper office, and I was given the privilege of attending at night. Joe Ropollo became my mentor. He was a principal reporter for the *Shreveport Times*. He let me go with him everywhere from a horse exhibition to trials. The editorial pages of the *Shreveport Times* are well known throughout the State. It is hard for me to describe the impact of being on the scene of any episodes affecting the State. It is nice to say that those years had a profound impact upon my military career.

Barksdale Air Force Base was across the river in Bossier City. I attended its opening and made a number of contacts throughout the years. One of these was with Major David N. W. Grant.

About this time, my Aunt Minnie married a fellow named Stuart Newland. Mr. Newland was a civil and sanitary engineer. Minnie Weil became Minnie Newland. Mr. Newland worked for a time on his own as a consulting engineer, and then as times hardened, he was an employee of the Work Progress Administration. He later went to work for the State of Louisiana and later Wallis and Tiernan Chlorinating Company. I personally walked the ditches of Shreveport carrying a rod for him and learned to make the plots of the city's drainage system. Stuart taught me a number of things including the basics of air conditioning and other important and necessary systems.

Until now, this has been a tale of very small events affecting the lives of very few people. Shortly after going to VMI, things began to change.

Chapter III

Life in Shreveport during 1937–41 included the oil discoveries in Haynesville Oil Fields and the growth of the air base. I graduated from C. E. Byrd High School in the Spring of 1937. On September 3, 1937, I matriculated at Virginia Military Institute. The next four years were very interesting and many memorable adventures occurred. Among these were the following:

The death of General Lejeune, who was at that time the Superintendent of VMI. General Lejeune was Commandant of the Marine Corps and had a distinguished career. Towards the end of that year, he passed away. A full dress parade and guard of honor was given to him. Following his death, the new Superintendent was Kilbourne. He was a stately man, a great leader and a wonderful friend.

When I returned to VMI for the start of the second year, I became acquainted with Colonel John Carter Haines. Carter Haines was a major influence on my cadet career and on my entire life later on. He was the adjunct professor of Civil Engineering and Sanitary Engineering. Carter Haines was the perfect embodiment of everything I was to become later on. I had already made up my mind to study sanitary engineering and to go on to Johns Hopkins. As a result of the Depression, VMI was a participant in Junior Work Progress Administration. I had sufficient funds to pay for my education, but many other cadets did not. They worked for WPA. I became a cadet technician at an early age and throughout my career at VMI, I was a practicing professional. Carter Haines was my mentor all through my career at VMI.

Among my notable achievements, which I undertook at that time, were the following: (a) studies on the use of a different media other than the traditional one for water; (b) use of cold water detergents and chlorine for sanitation; (c) numerous other investigations. These episodes were continuous throughout the following years at VMI. I received a letter from Superintendent General Kilbourne thanking me for the work I had done with regard to the flu epidemic in 1940.

Another notable event was the cleaning and painting of the columns of the Washington and Lee University.

I had the privilege of rooming with Seth Hobart and Hugh Davison. Seth was a pre-med student and Hugh was a liberal arts major. We stayed together for the entire four years. You really get to know your roommates almost like brothers in a four-year period. Also during this time, I met and grew to admire very much the Vice-President of the class, Jimmy Dale, and others too numerous to mention. Needless to say, four years at VMI were the happiest years of all.

Here is a summary of those years.

The Spring Hike. Each year in spring, there was a three-day trip that was made to a new high sight.

The VMI/VPI football games. Traditionally, these were hard-fought games and have long since become historic. VMI and VPI had tied 2–2 in a blinding snowstorm in my freshman year. The series had ended at 14–0. VMI had won in our senior year.

Another memorable event was the Spring Fling. I forget who my date was but it was a memorable event.

About this time in my life, girls became important. I had a series of adventures with Jeanette Sentel and others. I had a long-standing "affair" with a girl at Hollins, nothing serious but just a long-time standing, hoping to get married when we graduated. Once she had a fall from a horse, and I ran the block and rushed up to see her. I wound up with two weeks' confinement as a result of my impetuousness.

All in all, my four years at VMI were happy and good. I learned a lot about life and relationships, but most important, the ability to get along with others.

I do not remember who my Ring Figure date was. During finals, I had a delayed invitation from Miss Pie Ewing. My first choice for a date was Robin Atkinson, who had been injured at Hollins and therefore couldn't make it. Pie was sort of an afterthought but turned out very well.

Following graduation, there was a happy but bittersweet atmosphere because of the coming war. We were excited to graduate but had mixed emotions about the upcoming future of our country. This was sure to change our lives greatly. The war clouds in Europe were threatening and we did not know what was to come at Pearl Harbor.

The year 1839 was important for two reasons. First, it was the foundation of VMI on November 11, 1839, and secondly, the hundredth anniversary of the founding of Shreveport, Louisiana, by Captain Henry Miller Shreve. Captain Shreve led a group of men north on the Red River to attack the Red River Raft. Shreveport has always had a reputation, unlike south Louisiana, of being close to Texas. The streets in downtown Shreveport are Texas, Milan, Crocket, Travis and Fannin. The names of the heroes of the Alamo are well known in Shreveport. In fact, Shreveport is more like a Texas city than a south Louisiana metropolis. Being located thirty miles from Milan, Texas and forty-five miles south of Arkansas, the city bears little resemblance to south Louisiana. In 1937, the city still bore little resemblance to the rest of the State.

In 1938, my grandfather Harry Weil died. This was a tragedy for me. I missed him very much. The Superintendent called me into his office and told me that my grandfather had passed away and that the family felt that I should stay at school. I did but it was very difficult for me.

In 1939, I was emancipated. It made little difference to me at the time because my Uncle Tom was capably running the affairs of the Corporation. There were no real significant events after that until I left in 1941 to go into the Army.

The significance of being a grandson of rich men (Abe Meyer and Harry Weil) had not sunk in with me. All that emancipation meant to me was that I now had to sign some papers every six months. We held no Board of Directors' meetings or anything else requiring any action on my part other than to review the annual reports and submit any comments I had to my uncle. It was rather interesting to note that the bills as summarized were mostly for VMI on my account. In 1940–41, $900.00 was charged to A. F. Meyer. These fees were for tuition, room and board, uniform fees and other associated costs. While the rest of the world was still in a Depression, things had begun to improve in north Louisiana – thanks to the oil boom. Therefore, I was an unusual cadet because I was acting in my own right, although my interests were being handled by my uncle. It was becoming increasingly apparent that the events taking place overseas were going to have a major impact on the Institute and myself.

On September 3, 1939, Hitler invaded Poland and the events that started WWII were well known.

Chapter IV

Following graduation from VMI in 1941, I went with my mother and grandmother, Carrie Weil, to New York City and back to Shreveport. While I was not yet 21, it was at that time that I received my inheritance. My plans to enter Johns Hopkins were temporarily delayed by these orders. As soon as I was 21, I was to report to Fort Bragg. I was an entirely different young man in 1941 than I had been in 1939 and 1940. I began working at the Caddo Shreveport Health Unit as a young engineer, not as a technician. I knew I was going on active duty in September, but I took the job with the State anyway.

My years at VMI had laid the groundwork for a career in engineering and public health. I was ordered to active duty as a second lieutenant of field artillery. My request for transfer to the sanitary corps was turned down by the chief of field artillery. I reported to Headquarters of the 4th Field Artillery Training Regiment, and I was assigned to Battery A. The commanding officer was a first lieutenant and a VMI graduate, who was glad to welcome another VMI graduate. Our duty hours were relatively short and we had Wednesday afternoon off.

I coached the Battery boxing team because I had learned to box at VMI. I met a lieutenant in the Nurse Corps at a boxing match and took her on a horseback ride on December 7, 1941. We rode into the hinterlands around Fort Bragg. She asked me point-blank if I was interested in girls or men. I told her I had a steady girlfriend in Louisiana. She accepted that. We were coming back from the horseback ride and turned on the radio and learned that Pearl Harbor had been attacked. I turned it off thinking it was just another radio show. She said, "No, we should listen to this, maybe there is something to it." We listened, and it began to dawn on us that the world was at war.

The next morning, I attended the Battery formation and listened to President Roosevelt. There was a great change in the activity at Fort Bragg. I was promoted to first lieutenant after only three months of active duty and ordered to Fort Jackson in South Carolina.

At Fort Jackson, I was given the job of Battery Commander of Battery B 306 Field Artillery. I had myself as Battery Commander, three lieutenants from Princeton University and ten Army sergeants. We received our fill-in of two hundred personnel within a short while. These included draftees and volunteers from New York City and the surrounding area. Our job was to hammer them into shape.

I received an award as the best in the 77th Infantry Division Field Artillery in an annual shoot-off. I was transferred from my Battery Commander's job to Division Artillery S-4. Subsequently, I was sent from Fort Jackson to Fort Leavenworth in Kansas to attend the 12th General Staff Course.

Meanwhile, I had married Vivian Burford. When I had arrived at Fort Jackson, there was an excess of men over women, in incredible numbers. I sat down one day and wrote a letter to the Zeta Tau Alpha Sorority. In that letter, I stated that I was a young, recently made first lieutenant and a graduate of VMI, and I knew that the Zeta Sorority had the best girls on campus. I asked for a date. Sending the letter by special delivery, I did not expect to hear from them as fast as I did. Within a week, I got a reply stating that I had a date with Miss Burford, the secretary of the sorority chapter. Thinking of all the possibilities, the following Wednesday I dressed in boots and spurs and went to see Miss Burford. I parked by the entrance to the sorority house and was about to ring the doorbell when I looked up and saw this beautiful young woman dressed in nothing but a bra. Within a few minutes, she came down, and from then on we began to date.

We were married on June 13, 1942. We traveled to Fort Sill, Oklahoma. There we stayed in a hotel with several other officers and their wives. At Fort Sill, we fired live ammunition. We moved to the Louisiana maneuver area and then on out to the West

Coast. I had the opportunity to attend the 12th General Staff Course at Fort Leavenworth. When I arrived there, I was a Captain. I was told by the receiving officer that I was not a student, but that I was to be a faculty representative. I refused, saying that I was supposed to attend the course. After a discussion, it was verified that I was to attend the course. I graduated in June and proceeded on to the West Coast.

While at Fort Leavenworth, I met many officers with whom I continued to keep in contact later. Among them were Colonel Gus Nece, Colonel Ollie Niess, and Colonel William Patient of the Medical Corps.

The Division Artillery had already left to go overseas when I was in Phoenix. My son, Alvin Felix Meyer III, was born there. Through a series of misadventures, I wound up at Camp Roberts, California, as a medical inspector. While there, I received my orders to report to Carlysle Barracks for instruction and then on to Kearns, Utah. I drove night and day across country to arrive at Carlysle Barracks right on time. The series of misadventures continued because the orders were two weeks old and caused me to report two weeks early.

I used this period of time to look for a place for Vivian, Felix and myself to live. I found a delightful spot at Mount Olive Springs. It was a second-story, split-level resort. It was an ideal place because we could go downstairs and party while the baby was asleep upstairs.

After the two-month course, I was well equipped (I thought) to deal with the problems of a new commission to the sanitary corps. Everything I had learned at VMI fell into place with the new rules and regulations and information given at the advanced course. Vivian and I took the two weeks we had to go to Shreveport and then left with Gaylin, the daughter of Mary, my nurse in Shreveport. We were fully loaded when we left Shreveport and went on to Grand Junction, Colorado, and then on to Salt Lake City, Utah. When we arrived in Salt Lake City, I again ran into problems with the colored help. It was only after I made my point that I was allowed to stay at the Hotel Utah. We found a place that was for sale in the Sugar House District. I began to enjoy life as a sanitary corps officer.

There were a number of friends from VMI in AFORD (Air Force Overseas Replacement Depot) Kearns, Utah. Nash Strudrick was one. We also had a number of other people. Among them was a 21st Bomber Command Surgeons officer Major Ken Pletcher, and I had many conversations with Ken during the short time they were there. In the meantime, the atomic bomb was detonated at Hiroshima and Nagasaki. The 21st Medical Corps never left the State intact, but meanwhile I had been selected to come into that group as a sanitary engineer. When the war was over, I had a decision to make, whether to go on to Johns Hopkins and obtain my masters degree in sanitary engineering or to stay in the service. I decided to volunteer for the regular Army and was accepted into the Pharmacy Corps.

My daughter Carolyn was born in Salt Lake City, Utah.

Being accepted for the regular Army, I was transferred to Germany. Vivian took Felix and Carolyn to her mother's place and left them there while she joined me in the overseas replacement depot in Camp Kilmer in New York. There I was given the command of an attachment of Air Force officers. I explained to the camp superintendent that I was a medical service officer and could not command anything other than medical service troops. He said, "Well, you are a VMI graduate and the only one we've got right now, you have to take them." I approached the group and said, "Listen, I am a sanitary corps engineer and medical service officer, but I am also a VMI graduate. You will take my orders and follow them or I will see to it that you do not go any further!" On July 4, we sailed from Staten Island, aboard a liberty ship, bound for Europe. There was a contingent of infantry aboard the ship. I met the Lieutenant Colonel, commanding officer of infantry. We had a pleasant voyage over.

After we arrived in Bremerhaven, we disembarked and boarded a train. That morning, without warning, the train left, leaving our bags behind. After some delay, we eventually backed and recovered our baggage. The next morning, we arrived in

Frankfurt. These were hectic times in the European Theater. I went around to the mess hall and discovered we couldn't eat because we didn't have any orders. Feeling desperate, I found my way to a telephone and called up Lieutenant General Withers A. Burres, who had been Commandant of VMI for two years when I was there. I explained who I was and he said, "Come on up here." So I did. Shortly thereafter, we got the troops fed and on their way to First and Felbruck. I arrived and turned over the command of the unit I brought and was told that I could stay. I told them I had orders to be at the headquarters of USAFE (US Air Force in Europe). Finally, after numerous phone calls to USAFE, Colonel Wilford Hall came down to fetch me.

I was told that I had been on orders to Wiesbaden, Germany, for three months. The first thing that I was told I had to do was to get water into the city. I was introduced to the commanding general who said, "Your job is to get water into this city, I am tired of washing my teeth with champagne." The city's water supply was an interesting one. It involved three levels. One upper level was from the Taunus Mountains, the third level was from a steam pumping station on the River Rhine. The mid-level was controlled by a reservoir in which the two other levels met. During the allied bombing raids, the city of Wiesbaden was virtually unaffected except that the raids broke the connections of the water system, and the system was virtually empty of water.

Using my usual ability to find solutions, I began to investigate the system. Wiesbaden had a population of about 250,000. It was one of the few cities in Germany that was undamaged. A large number of residents were displaced by people fleeing other cities. Water became a major concern of the allied occupation forces. I began to try to find a solution. I went to the Taunus Mountains in the infiltration galleries and into the steam power plant on the Rhine. It became obvious that the solution was one of the (then) modern systems used in large American cities. Fortunately, I found a large water supply plant in the explosives department of a paraffin plant on the Rhine. I then began a series of important events.

I traveled to Frankfurt to US Forces Europe Headquarters and talked to Mr. Potts, the sanitary engineer, about getting pumps. I located a large number of pumps about to be sent to Russia. To make a long story short, I liberated the pumps and made the necessary connections. We soon had enough water to flush the toilets and maintain the supply throughout the city. Because of my efforts, I received the Air Force Commendation Medal, the Army Commendation Medal, and from the city of Wiesbaden a small plaque showing a mother lifting water up for three small children.

After taking care of the city of Wiesbaden, I was then able to take care of a number of other items. Vivian finally brought Felix and Carolyn over, and we began to live a normal family life.

The Rhine main airfield was about to be the focal point of the Berlin airlift. It was a monumental effort to transform the damaged airfield into a tent city and then into a permanent city. We improved the airfield with very many tons of supplies. I was called upon to devise and improve a sewage treatment plant. The original Imhoff tank had been designed by Karl Imhoff himself. Without going into details, I met Dr. Imhoff, and we figured out a way to convert the present system into a modern sewage system. At that time, the airfield began to fill up with Swiss huts, and other facilities like the water and sewage disposal system were serviced, like none other at that time.

There were many other things that were happening throughout the theater. One big operation which we commanded was the aerial spray of insecticides. The German Army in the Crimea retreated north and west reaching Berlin. The defenders of Berlin were afflicted with malaria. Following the devastation at Berlin, the Anopheles mosquitoes became prevalent and affected the Russians and US forces. My boss, General Schreuder, had visited Orlando, Florida, at the same time as I did in 1943. He had the plans for aerial spray. We immediately took those plans and used a plan that would utilize a C-47 and three L-5s. When we first approached the Russians about spraying in Berlin, they were antagonistic. We told them we would spray anyway: if they wanted to let their troops get malaria they could, but we would help ours. They finally agreed and we sprayed DDT by air over

Berlin. We successfully broke the back of the epidemic. Not only that, after we sprayed Berlin, the Austrians requested the spray planes be used to combat the Dutch Elm disease, which was affecting the forests in that country. Russia again demurred at first, but later agreed.

Meanwhile, in North Africa, a serious cholera epidemic was beginning to emerge. The presence of flies was the contributing factor. I have a picture of a flight of airplanes spraying DDT over the pyramids.

General Schreuder left to go to the Command Surgeons Office, and shortly thereafter, I was transferred to Dayton, Ohio.

Chapter V

We arrived at Wright Patterson Air Force Base, along Highway 4, in 1949. It was just the same as it is today. Starting in Dayton, it goes along Highway 4 right past Wright Patterson Field and the flying bases. It is an exciting place. It was more exciting then than it is now.

After some delays, I found a house on 16 South Street. It was along the base fence to Area C. The New York Central Railroad ran past it and it was a very lovely place. It had a huge backyard. It took me approximately two years to fix up the house the way I wanted it.

Otis Schreuder was my boss, as he was in Germany. Bob Peterson was the industrial hygiene engineer. He was a very fine fellow. John Boyson, Deputy Surgeon, was an excellent mentor and trusted friend. In the intervening years, we discovered a number of things together that we enjoyed. John and I became very close friends. Together, we wrote a paper entitled *The USAF Physician Engineer Team of the Future*. We had it published in the *Military Surgeon* that year.

There were many things going on that I discovered I had to tackle in the job. Among them was an epidemic of infant blue babies in the United States.

We were faced with a problem of disposal of red fuming nitric acid. This missile propellant was a feature of the then emerging technology. However, red fuming nitric acid at Wright Patterson was contaminated with water. We tried to dispose of the material in various places. Finally, we went to the State Health Department in Columbus and sought their advice. We told them we planned to get rid of it. They proposed that we dispose of it with a dilution in the Mad River. Following that, we went to the US Public Health Service in Cincinnati and talked to them, and they agreed with the idea of dilution in the Mad River as a means of disposal, provided we could control it by laboratory testing. Then, we went into the laboratories at Wright Patterson to discover any personnel who had experience of fish. We discovered in the laboratories at the hospital one individual who was a fish expert. We also found a number of other personnel. General Schreuder was willing to forego spaces and use of people for a laboratory. We started the Air Force Occupational and Environmental Laboratory with only the authorization of the Air Force other than our own willingness.

The Aero Medical Laboratory at Wright Patterson had a surplus electron microscope, which they donated to us. We used it for the first industrial hygiene section. We found space at an old hospital building in the old Wright Patterson Field Hospital. Peterson took charge of the lab and I was able to manage the red fuming nitric acid.

Before we released the material, we (I and Gordon MacEachern) took a boat ride in an aluminum canoe, which I had, down the Mad River to Dayton. We verified that the Dayton water supply came from the Mad River. In spite of that, however, we decided to go ahead with the disposal.

Another very important episode was the consultation program for the industrial hygiene and sanitary engineering for the bases. At that time, it was myself, Bob Peterson, and a number of engineers, before the Korean War. The Air Surgeon at USAF Headquarters had given the Surgeon AMC responsibility to provide consultancy services to the rest of the Air Force. We decided to formalize this with Air Force regulation. After many delays, we published the regulation, which was a fact with which we could provide a very limited form of industrial hygiene and sanitary services for the Air Force. Later, we were able to put in more of an emphasis that we obtained in the Korean War and more troops. One of the early additions was Colonel, now retired, Herbert Bell. Herbie and wife Nancy became very close friends. It was during this time that I asked him to learn more about noise. Herbie told me he was a civil engineer who didn't know anything about noise. I took a moment and then asked him if he remembered taking a course in thermal dynamics at VMI. He said, "Yes." I said, "Didn't you study noise and sound?" "Well yes," he said. "Well, noise is unwanted sound," I said.

General Schreuder was a WWI aviator. He liked to take off in a B-25 and fly to various places. He thought nothing of leaving at 4:00 A.M. going to Kelly Air Force Base and leaving there at 5:00 or 6:00 P.M. for a four-hour flight to Wright Patterson. He was also a fine gentleman and was always comfortable with his staff. General Schreuder was a competent industrial medical officer who had a number of fine consultants. These included Dr. Robert Kehoe from the University of Cincinnati, who led us in a number of adventures. Another of Schreuder's consultants was Dr. Henry Vaughn, an engineer who was head of the School of Public Health at the University of Michigan. These individuals helped form the background to the rest of my career.

The Korean War brought a vast expansion of responsibilities for the medical service. We began to recruit industrial hygiene engineers, aggressively. In 1951, we got the first large group of industrial hygiene and industrial sanitary engineers. We thought at first we would train the new engineers at Wright Patterson. However, we didn't think that we had to transfer them down to the Medical Service School which was at Gunter.

Along 16 South Street were the following: my next-door neighbor was Lynn Weret, a test pilot. Across the street, directly opposite, was an individual who would soon electrify the world with his exploits, Chuck Yeager. Down the street in the next house was a VMI colleague. In the next house was Colonel Richard Cole and his wife. He was in the raid on Tokyo, led by Colonel Jimmy Doolittle.

All in all, it was a small fragment of a larger issue. We had a wonderful time socially on Saturday evenings. I will never forget Lynn Weret and his Lincoln, which was parked alongside our kitchen. Lynn was a great practical joker and played a wild trick on us. He called one night while I was on a temporary duty elsewhere at about 10:00 or 11:00 P.M. and told Vivian, "At last, I know you sleep naked!"

About this time, we had also inherited a dog named Rebel. Rebel was a very uproarious young English cocker spaniel. One night while we were at the officers' club, Rebel had managed to open the kitchen cabinets and spray the soap flakes all over the kitchen. Lynn threw a potato at me, shaking his head. I said Rebel had to go to obedience training, after which he was a perfect dog.

Among those whom we will see later on are Clarence Feightner and Frank Smith. Feightner was an industrial hygiene engineer at Tinker. Bill Fluck was also an industrial hygiene engineer at Sacramento. Ed Poth was an industrial hygiene engineer at Kelly. John Pearce was the one whom I had known while I was at Kearns. In fact, the three gentlemen were all we had on active duty or that were available as civilians during the immediate post-war period.

At the start of the Korean War, we began active recruiting. We needed to obtain the services of several engineers. We found Herbie Bell, Lynn Channel and Frank Smith (not the same Smith from Sacramento). Bob Peterson proved invaluable to me at AFMC Headquarters. As the war progressed, Peterson came on active duty. Feightner and Bill Fluck were also called to active duty.

Chapter VI

General Schreuder retired, as was required by law, and was replaced by General Tracy. Tracy was an elegant gentleman but also had a heavy reliance on alcohol. He had served with General Schreuder as an officer overseas during WWII. My transfer to SAC Headquarters was announced by the Air Force. He took me out to SAC Headquarters. I met Lloyd Griffis, who had been Surgeon of the European Transport Command during the Airlift. I had known Griffis over there and found him to be exceptional as a leader. I was to replace the existing SAC Command's sanitary engineer.

OFFUT Air Force Base

After a long summer's ride, we searched for a new house. I found a house for rent near the Aksarben Race Track. We lived there for one year until we built a new house at South 94th Street.

Before I became active in SAC, I joined St. Andrew's Church. We met Jane Freeman and Max her husband (executive at Procter & Gamble). Also, we met several others. You will meet Max and Jane later.

At SAC

An early encounter, where I shall mention the role of Jack McCambridge. McCambridge was a young officer I met at a visit to Fairchild Air Force Base. I was very impressed with Jack. That night at the officers' club he told me everything that was wrong with the Air Force and that they could make much more money as engineers in Seattle. I listened and thought to myself that *this guy needed to learn a lot but that he had a lot of good points*. Shortly thereafter on my return to SAC, I had set up an important interview with him at Headquarters. He thought he was coming in to catch hell but instead, much to his surprise, I offered him a job. He proved to be a very valuable officer, and later was assigned to the Pacific Missile Range, now known as Vandenburg.

He did a magnificent job at Vandenburg with Colonel Raymond Yerg. I will never forget the night I visited him out there and in the fog we returned to the quarters with a young dietician, to whom he was engaged to be married.

Over the next several years, days were filled with very exciting things. I saw the move from underground bomber plant to a brand new underground command post. Ted Bedwell was a Colonel and became Command Surgeon. During Bedwell's time, we had many exciting events, one of which was the dropping of a bomb on a farm in South Carolina. According to the Aircraft Investigation Board, that event occurred accidentally. The crew of the B-47 had a warning light in one of the bomb bays. The aircraft commander sent the flight navigator back to see what was wrong. According to the report, he put his hand on the bomb and the bomb fell out of the aircraft. The officer was almost sucked out of the plane with it. This weapon was one in which the firing device was kept separate. I was on leave, and Ted and Herbie Bell flew down with the disaster control team. The major damage from the bomb was the destruction of the farmhouse. The following result was a lawsuit. This was the first "broken arrow" which I had any information on. I took the team, including Hugh Mitchell, down to Sumpter, South Carolina. We successfully defended nine stakes against claims for damages. After that, there were several disasters affecting a "broken arrow".

The most memorable event was a collision of a KC-135 and a B-52 over Fort Knox, Kentucky. I was attending a PTA meeting at Westside High School. My son, Felix, was at home and took the message from SAC Headquarters. I left the meeting, changed clothes, and reported for duty. En route, General Ryan, "broken arrow" Commander, told us that very likely a nuclear explosion had occurred in the air. The sky lit up from Fort Knox to Cincinnati. Due to the fact that bombs are almost impossible to defuse, I felt that a nuclear explosion was not likely. At the airfield adjacent to Fort Knox, we were met by the Army commander who expressed great relief. We went to the site of the crash of the B-52. The KC-135 had landed in a field nearby. There were no survivors in either aircraft. After a lengthy search, we discovered one of the bombs in the wreckage did not fire. General Ryan and I handled the fire and flames, and a member of the Explosive Ordinance Disposal Team verified that the weapon was intact.

Another episode involving nuclear weapons was at a southern airbase. A B-52 that was approaching Hunter Air Force Base caught fire. The aircraft commander ordered the crew to separate them from the airplane. The bombs fell. When we arrived, I will never forget the sight. The four bombs were hanging from parachutes in a tree. Knowing that the activation of the weapons was dependent on the force of the fall, we did not know if they were armed or not. We approached them gingerly to discover all but one of them intact. This one had split open, which began a long search.

Military life was not all excitement like "broken arrow" incidents. Instead, there were many nights of routine aspects of occupational medical service and other things affecting the health and safety of the aircrews.

Maintenance Personnel

We worked long and hard at removing the causes of non-effectiveness in SAC.

During the third time of my stay at SAC Headquarters, I felt that this was the most exciting time of my life. General Power was a different personality from General LeMay. LeMay was a rough, hard-driving commander. Power was a gentleman, scholar and a great leader. I did not know General LeMay very well, but I got to know and respect General Power very well.

During this period, the Joint Strategic Target Planning Agency was created. The Navy Vice-Admiral was ordered to SAC in Second Command. General Blanchard and I were members of a team which participated in war games. General Power gave us the job of putting out of action our own aircraft carriers. We devised a method, which we used a number of times attacking aircraft carriers. The high point of the details list was when we successfully launched several strikes against our own aircraft carriers. I obtained a model of an American nuclear submarine and painted on the warheads and missiles a SAC insignia. We left it on the Vice-Admiral's desk. He did not take kindly to this. I was promoted to Colonel, becoming one of the youngest colonels in the medical service. I was ordered to the National Industrial College of the Armed Forces. General Power said I was too valuable for SAC to go to the College. His petition to General LeMay was, however, denied.

Although I had a very busy life, I had an active, social and happy life. The University of Nebraska football team was a great event in my social life. We belonged to the Omaha Athletic Club and took the train to Lincoln for every game. Felix and Carolyn were growing up rapidly. Felix wanted a car and I told him that he would have one if he earned half the money. If he did that, I would match the other half. He went to work at McDonalds in the early days and rose steadily to being a full-time manager. Carolyn was badly burned one year and despite the burns she went to work as a lifeguard. Vivian and her friend Jane were busy with St. Andrew's Church in which I was a Junior Warden and later the Senior Warden. All in all, it was a very happy time. We had numerous peacetime engagements of social types, both at the base and other local country club scenes.

Van Chamberlain, the Director of St. Andrew's Church, was a very important player on the golf course as well as in preaching.

Industrial College

I received profound assurances from several officers that I would be returned to SAC when I completed the Industrial College. Therefore, I leased a house for one year and went to Washington. To my dismay, in the fall I received orders to continue in Washington upon completion of the Industrial College. Meanwhile, I have several events to tell you about.

First and foremost, my daughter Carolyn ran away and I spent several weeks looking for her. Finally, she was located and we brought her home. She got married.

The change from the reserve to active duty status of SAC to the peacetime situation at the Industrial College was quite an adjustment. No longer did I sleep with the red phone next to my bedside. Nor did I find myself on the go. The Industrial College was a fine experience. One of my closest friends was David Crocket, a naval aviator. We played golf together nearly every day during the winter of 1960 and 1961.

Knowing the set-up of the Surgeon General's Office, I began to make plans for leaving the service in four years. Vivian was very unhappy and I was too. We figured that we could stand four more years and try to get an assignment because I only had twenty-three years. After that, I was going to retire from the service. In four years, however, we took a South American cruise along with several of our classmates. We went to Buenos Aires, Sao Paulo and other cities in Brazil and Uruguay. It was a notable trip, full of surprises, and very good. Before I graduated, I received from General Power the first Missile Badge awarded to a non-missile officer, as well as a commendation for my services at SAC.

Upon completion of ICAF, I was to report to USAF Head-quarters. At that time, the Surgeon General's Office was located in T8, an old, rambling wooden structure, about a mile and a half inside the District of Columbia, beyond the Washington Cathedral. The Surgeon General's Office occupied the second floor, whereas the Military Sealift Command Office of Chief of Chaplins USAF was on the first, along with a cafeteria. The building was depressing enough. I was welcomed by General Bohannon and several others and took my place in the Aerospace Medicine Division. My original feelings were that I was going to make the best of it for four years and then get out. The first five or six months were just as I expected them to be, nothing much to do and very little else to my assignment. However, shortly thereafter, I began to make some progress on various jobs. Some of the aero staff had an attitude of hatred towards congressionals. I soon became aware of the fact that there was not a great deal that one could do for the Air Force in that realm. By the end of my first year, I began to be very active in doing certain things for the general good of the Air Force. These were the hearings on the missiles' effect on the missile propellants on air pollution. The toxicologists had identified that beryllium being used as a rocket propellant had a tremendous effect on human life. A long time ago, we had organized teams to investigate the problems of beryllium and rocket propellants. Among the results were findings that no significant hazard existed, when Owen Kittilstadt had run a series of tests at Edwards Air Force Base for us. We assembled a team and gave a series of hearings that showed that the Air Force was well ahead of the rest of the country on this problem. In rapid succession, a series of hearings on missile propellants on the Navy and Air Force and operations in conjunction with their work were held. Around the second year, I was well engrossed in various projects of this nature.

General Bohannon had as his deputy Major General Ken Fletcher for a while and then General Towner. I had known Ken when he was a major. The next real opportunity came following a report of a committee on the incorporation of the Medical Services into the DOD (Department of Defense). This report was successfully charged by the Air Force, then the Army and Navy. I played a major role in writing the Surgeon General's defense. As a consequence of that report, it came out that something had to be done about the allied health professionals. The Air Force, in particular, was having environmental engineers. General Bohannon said to me that he wanted a comprehensive study made by the group and that I was the one to make it because of my ICAF training. The problem was in the Air Force, in particular, as well as the Army and Navy, that in order to be promoted to full Colonel, a Medical Service Corps officer had to switch over into the supply and administrative fields. Not many of us were willing to do that. The one exception was that the bio-environmental engineers had already achieved some independence. After several months, I wrote a report recommending the formation of a new corps in which its chief would be selected from among the ranks and also the associated chiefs, heading each of the specialties. In addition to the optometrists, pharmacists, clinical laboratory and other specialties, as well as bio-environmental engineers, and, of course, the Medical and Women Medical Specialists, the corps should contain dieticians, physical therapists, and occupational therapists. This latter group comprised a very active opposition for the new corps. I finally met with the leaders and said, "Do you want to be a full Colonel or do you want to be just Majors and Lieutenant Colonels?" At that time, the act forming the Women's Specialists Corps consisted of a full Colonel from the three corps and three Lieutenant Colonels comprising the rest of them. We had Majors, Captains, and Lieutenants. I pointed out to them that this was not fair and that all ought to have an opportunity to compete across the board. The difference from the Army and Navy was that the Secretary of the Air Force could comprise a corps if he wanted one, whereas the others required legislation. Finally, the Secretary made a decision that the Biomedical Science Corps would be formed. It is now the third largest corps of the US Air Force Medical Service.

Along with the rest of the staff, I went to the dedication of the new Aerospace Medical Division Complex at Brooks. The President of the United States, John F. Kennedy, gave an inspiring speech. We left the next day to fly back. While we were in the air, we heard about the President being shot. We landed at Andrews Air Force Base and hurried home. We did not know that President Kennedy was dead at that time. It was a great tragedy and loss for the whole nation.

While on active duty, I was appointed to a committee in the White House. It was chaired by John Buckley. The inter-agency committee was a high-level working group on environment. It existed in the early years of the Nixon Administration. When the EPA was created and the Council on Environmental Quality came into being, the committee was abolished. Part of that was arranged for all the problems facing the nation and was reviewed by the committee. Among the accomplishments that I was involved in were: (a) Project Chase – the Navy had established a program of cutting holes and sinking the ships. This involved transporting across the country massive amounts of explosives to one or two places in the United States. They would load and board the ship and take it out to sea and cut holes in the ship and sink them. The affair became a national issue when a ship was lost at sea before it was actually cut and sunk. It was a near disaster and scarcely averted. Chairman Buckley and I went out to the base in New Jersey at the time of a launching of a cut-holes-and-sink-them expedition. We verified that ships would not sink below a certain depth. As a result, cut-holes-and-sink-them was abandoned. (b) Another project for which I was principal was a high-level committee to derive a new expression whereby Defense would reduce pollution from its contractors. I headed a group of high-level officials organized into several groups exploring different approaches to pollution prevention. We finally agreed that the only hope and practical way was to control pollution as an element of contracting. As a result, the Federal Acquisition Regulations were changed, making it necessary to evaluate whether or not the 300,000 federal contracts were indeed conforming to the Air and Water Pollution Act. This result of over a year's investigation was all we could do. It still represents a major step forward in time.

One of the very interesting things during my experiences at USAF Headquarters and the Surgeon General's Office was the relationship with Congressman Jones, head of the House Investigative Committee, and Senator Muskee, Chairman of the Air and Water Pollution Sub-Committee of the Committee on Public Works.

My relationship with Senator Muskee began as early as my second year at the Surgeon General's Office, and continued throughout the entire service of the Surgeon General's Office and later as an employee of HEW.

I testified many times before the Senate Sub-committee of which Muskee was the chairperson. Defense was the "whipping boy" of the early days of the EPA. Muskee's principal advisor was Leon Billings. I had discovered that Leon was a graduate of MIT and knew Rolphe Eliasson. An uneasy truce developed between us at first – between me and the Chairman of Defenses Departments' Committee on Environmental Pollution Prevention. I was determined to make our points. Begrudgingly, Muskee began to agree with us. We opened a new laboratory for the environmental pollution prevention activities of the Defense Department (notably the Air Force) and invited Senator Muskee to go to the west coast to inaugurate the laboratory. He accepted and we left. As I remember it, we departed from Washington aboard an Air Force airplane the evening of his opening campaign for President of the United States. I learned a lot from Senator Muskee and Congressman Bob Jones.

With regard to Congressman Jones, I accompanied him on several trips. He held a series of field investigations of the Defense Department. He said to me in the early hours of our first trip that he would make an attack upon me and the rest of the DOD but not to take it to heart, because when the meetings were over, we would still be friends. I understood perfectly and every time he made an attack upon the DOD or the DOD Installation, I took it calmly and later on we were still friends.

When I was contemplating leaving the Air Force and going with Health Education Welfare (HEW), I talked to Senator Muskee about this. He said it was a good idea and he hoped that I would go ahead and take the job that was offered to me as Legislative Liaison. Bob Jones gave me similar advice. Both individuals seemed to think that the Defense Department was ahead of the rest of the country and should be given credit although all their marks had otherwise indicated that Defense was the culprit.

As head of the Defense Departments' Environmental Committee, I had numerous occasions to defend the actions of the DOD as a culprit. Early in the turn of events was the attack on the DOD regarding water pollution. I remember one episode in particular. It was the contamination of the San Diego Bay. The EPA wanted us to bundle up the ships and immediately correct them. We took a team of experts to the west coast and looked over the situation. To make a long story short, we came back and provided a thoughtful explanation in which we stated that the Navy had many problems and we were attempting to solve them.

In addition to the problems with the Navy, the Army Corps of

Engineers had a larger floating plan than the Navy. Many of the locks and other pertinent systems dealing with inland waterways were the responsibility of the Corps of Engineers. We had an aggressive program started that continued even long after I was gone.

In each of the hearings, we had a team of individuals in uniform, including myself. Senator Muskee was always very receptive to having a full presentation of the defense of the position. Years later, when new ships were on the line and the older ones had been retrofitted, I felt that we had accomplished a great deal.

I had made up my mind to get out of the Air Force. I proceeded as calmly as I could to settle things. While I was in the process, I received a call from Surgeon General Bohannon. He wanted me to meet him at the visiting office of the Chief of Staff. I went home, changed into my uniform, and met him at the chief's office. He explained that our work on children had potential (CHAP – Children Have A Potential – Program). It was very important and he was forced to reconsider my position. He further said that the Vice-Chief of Staff J. P. Ryan wanted to talk to me for a few minutes about my retirement. I agreed to talk to him.

When I met J. P., he said, "Doc, I want to tell you that you were on the list but Bohannon had you taken off. He said that the Surgeon of TAC had to have a star and there was only one star and he was going to get it." This confirmed my suspicions that the words of a Surgeon General were worth nothing.

J. P. went on to say that he would be the next Chief of Staff and that if my name came up he would endorse it and see that I got the star next year. I told him that I appreciated his interest, but that I could not take the chance. J. P. and I had shared the "broken arrow" at SAC and had been on the "broken arrow" team at SAC and had many other occasions to work together. He said to me, "If I were in your shoes, I would do exactly what you are doing." I went ahead with my plans to retire and turned over to Fran Ballantine, my Deputy Chief, the reins of the Biomedical

Science Corps.

I was offered from the Public Health Service a three-star position as Director of Legislative Affairs. My boss was to be the chief of the newly created "Consumer Protection Environmental Health Services". I reported there immediately and was engaged at once in a number of many interesting events. These included the reconstitution of the acts of which the various supplements of CEFUS were operating. C. C. Johnson was the Assistant Surgeon General in charge of CEFUS. There was a long-running fight between the Director of the Air Pollution Division and C. C. Johnson. This extended for a long period of time and was not over until the EPA was formed.

Governor John Burns was a dynamic individual who had served his country in WWII. He later became Governor of the State of Hawaii. I took the opportunity to meet many friends in Hawaii, including the managing partner of a large construction company, who later became one of the Governor's staff.

I wrote the first executive order outlining the responsibilities of the new organization and also prepared the list of all the codes, standards and regulations, which had to be changed. I prepared a list of all the equipment that had to be purchased. The Governor wanted me to stay on after my assignment. However, I declined and said that I did not think I could be a part because the State had a requirement that you had to be in residence for ten years if you were not a native son. He said, "I don't care what the laws are. I will make a new one and have you installed." Well, I came back to the States after serving one year in Hawaii.

After my Hawaiian tour was over and I returned to the States, I became involved with the Office of Noise Abatement and Control. I was given the job of making a report to Congress and the President within the year. Central to the idea of the report was a concept of public hearing.

The newly formed EPA oversaw the Office of Noise Abatement and Control. I was appointed to head it and write a report to Congress on the state of noise as a public health problem. There was a great deal of confusion about noise. Many people felt it was a terrible problem and others tended to ignore it. We created panels to get views from both sides and participants from all parts of the community. We followed with technical experts, and the results on the noise pollution were divided up into a series of reports for the President and Congress. Recommended from our office was that the office should establish some new regulations. I will not go into any great length as to the results of this effort, but I will state that towards the end of my tenure as chief of the office others in the staff of the EPA decided they wanted the job. I quit. It is noteworthy that the Office of Noise Abatement and Control was abolished in later years and that regulation of noise devolved upon the states. That is where I thought it belonged in the first place.

Following my leaving the EPA, I set up my office in the basement of 1600 Longfellow Street in McLean, VA. I had long since kept an office there; having been presently involved with the dealings with the A. F. Meyer Corporation, I had this office area already. I spent a good deal of my time in Shreveport, Louisiana, running the A. F. Meyer Corporation and my other career. I was approached by Roy Weston asking me to come and take over the job of legislative liaison officer. I refused. However, Weston asked me to help put together a proposal for energy research and development administration. I did and we won the contract.

I then began a series of developments, which resulted in my concentrating my efforts on my new firm of A. F. Meyer & Associates, Inc. In addition to the work with Roy Weston for that operation, we also did work for them in the City of New York. Weston's assistant left him and went to work for Booz, Allen and Hamilton. We continued work with Roy Weston for several more years.

I became interested when opportunities were offered by the Navy. We won a contract with the Navy's Unit on Environment Support Activity. This was for a contract to go around to Navy installations and to give them advice on correcting their occupational, safety and health deficiencies. We were successful the first, second, third and fourth years. By that time, we were the only ones selected to continue the work. We visited some aiding installations and had a very successful effort. In conclusion, I told Paul Yarship, the Director of Civil Engineering in the Office of the Navy's Engineering Department, that we would not continue this work. I could not see us taking federal money for this effort that had been completed. However, we discovered there was a great need for hazardous materials. After some delays, a new contract was written and given to us.

This contract was for the modification of the Navy Occupational Safety and Health Manual embodying hazardous materials control. We visited some twenty-odd sites including major commands. After the conclusion of this effort, we became aware that there was more of a general requirement of logistics. I recommended that the contract be broadened and the manual be recast into a Navy Manual of Regulations.

The draft of the Navy Manual was changed into a regulations manual entitled *Hazardous Material and Control*. It was established that it would be in a different part of the Navy action. Looking back on it, what made it a logistics manual was that we put it in DMAT (Directorate of Materials). That was after its implementation.

Suffice it to say that the next five years we proceeded to participate in every aspect of the Navy's hazardous control and management program.

This chapter describes experiences we had with the boat, *Escape III*. What is described in the previous chapters occurred throughout the life of the boat. Because it is so important to me, I have written this as a separate portion of the book.

It picks up with the construction of *Escape I* and then goes on to describe in brief a long and deep involvement with boats and boating on the Chesapeake Bay.

I asked Max and Jane Freeman to join us for vacation. We took the *Escape I* and cruised all the way to the head of the bay. Going back, we ran into extremely heavy weather. We finally docked at General Talbot's. The next day we left, we had to rush to get back to Washington. Approximately one hour out, a disastrous explosion occurred and we wound up in the water waiting for rescue. A passing boat stopped by and picked us up. Max was severely hurt in the legs and I was burned about the body, but the girls were okay.

A later investigation revealed that the carburetor on the engine had separated from its mount and caused the explosion. I should have filed suit against Chris Craft but did not. Vivian and I continued boating as we bought another boat named *Escape II*. Then we bought a seagoing houseboat named the *Escape III*. We kept *Escape II* for two years. I can tell many stories about the times we had on the *Escape II*, but I will save those for another edition.

We kept the *Escape II* in a shed at the boatyard. We met a doctor who was a psychiatrist at the School of Medicine at the University of Maryland He and his wife are very close friends.

The large houseboat is the *Escape III*, thirty-eight feet long, eleven feet wide. She was built by the Seagoing Houseboat Company at Mussel Shoals, Alabama. I visited Mussels Shoals while the boat was being built and saw the hull. I was much impressed by the strength. Vivian did not want to buy the boat because she thought it looked ugly. I explained to her we were on the inside and not the outside looking at it so it was up to us to enjoy it. She finally agreed. We took possession of her at Fort Washington Marina. We took her around in the Potomac up to Point Lookout and then to Annapolis. The first problem we had was we discovered after about two or three weeks on the Chesapeake Bay that the deck house fittings were not all non-corrosive stainless steel. I wrote to the Company and told them I had ordered stainless steel screws. They shipped, after a day or two, some screws to us and were going to pay for having them installed. Instead, Vivian and I completely reset all the screws in the deck house, a long job for the summer.

Later on that summer, Olin Hester, Ed Menegeaux, and Vicky and Roger Stewart were our cruising companions the entire time we were on the boat. The next two years, we only escaped three times, and Vivian engaged in keeping her data in good shape while aboard. I intended to use *Escape* every weekend and we did.

After a summer of preparation, we started going out every weekend on the boat. The yacht yard was owned by Dick Vosbury at that time and was a nice place to do this.

We spent every weekend on the boat. It became a way of life with us, not only dockside, but cruising the Chesapeake Bay. We went from one end of it to the other. The *Escape* was really our escape. We visited many places aboard the *Escape I* and *II*. This time was more leisurely and great fun. Gradually, the *Escape* became a focal point to see all our friends. Olin Hester and Ed came aboard every night and every weekend. One weekend we had an experience. We were sitting on the couch up in the forward area and we were talking about where people came from. Olin said he was born in South Carolina in a small town nobody had ever heard of. I asked him if he and my wife knew where it was. He said "Calhoun Falls". I nearly fell over because it was Vivian's home. Later on that day, we discovered that Olin had left Calhoun at age about two or three and had gone to Miami with his brother. From that time on, Olin and I became fast friends. You will see him later in our discussion. The group became a cruising prom. Officially, it was called the Royal Creek Yacht Club. It had no rules and a lot of discussion on Saturdays and Sundays. We watched television and a period of ball games until the final episode of *Dallas*. Many frank discussions were held.

Among the other groups using the yacht club was Singles on Sails. Singles on Sails was mostly made up of women. Olin Hester did not belong to it, but Ed Menegeaux did. This group had a social happening every week. I did not go but Ed did and told us about it.

By now I had a contract with the Navy and was spending some time away from Washington and Annapolis. I was at the Navy's test ranges and higher test centers. When I came back, I was very dehydrated because it was so hot out there. I flew to Baltimore and came onto the boat. I woke up with a terrible sweat and chill, and the next morning, I went to the bathroom on the boat and voided blood. I called the Chief of Flight Medicine at the Pentagon, and he told me that I should urgently go to the Navy Hospital at Annapolis. I took the car by myself because

Vivian did not want to go, and went to the Navy Hospital. To make a long story short, I collapsed in the waiting room. When I came to, the doctor told me that I was seriously ill with a bladder infection. He told me I had two choices, either take the ambulance and go to the hospital in Annapolis or take a long route over to Andrews. I asked him, weakly, what he would do. He said that he would go to the hospital in Annapolis. I assented. Several funny things happened during my stay at Annapolis Hospital. One of them was I could not eat, and the dietician in the hospital came to see me. She wanted to know what was wrong. I told her I simply could not eat. I had this bladder infection and just could not eat. She said that it was just all in my mind. I told her if she served me something besides standard hospital food, perhaps I could.

After several weeks, the infection came under control and I was able to get up and take a shower. The first time I took a shower, I fainted again. This time, a nurse had come to help me. I held on to her and collapsed. She said later, "I have never been felt up before and you are surely really sick." I replied that I was and I did not recall any memory of her at all, much to my regret. However, I was well enough to go back to the *Escape*. Olin Hester came and picked me up. Vivian, in the meantime, had not been able to come to the hospital to see me. She was having trouble with her drinking again. After several days, I went to the Air Force doctors at the Pentagon, who examined me very carefully and said that I did not have a tumor in my bladder as they thought I had at Annapolis Hospital. After Vivian died, I had a short span of not going to the boat or doing anything, but then I married Susanne Sideman and continued boating, as in the past. After she was killed in an automobile accident, I married Grace Weil and continued boating.

The yard was sold by Dick to the Lilly Brothers. The two brothers were as different as the day is from night. One was very concerned and astute, while the other, Art, was a reckless character and was very likable. He owned a boat called the *Vitamin C*. It was kept right opposite the *Escape*. I have two subtle memories of the fine things about the *Escape* and several about the

Vitamin C. One of these embodied girls. Vivian and I were sitting abroad the *Escape* one day and noticed a girl coming down the dock and boarding the *Vitamin C.* After a while, Art came down and closed the cabin's drapes. When he left, the girl was there by herself. Another girl, meanwhile, came down the dock and came aboard the *Vitamin C.* There was a terrific fight between the two girls. Art, meanwhile, had gone. Another time, the girl and her mama went ashore in a small boat. They went over to the Hilton Hotel. Another girl came along and entered the boat with the first one, and Art left. Another time, it was quite different with him. We noticed two boats from the Coast Guard and Police Department come ashore. On our dock there were two policemen. Art was being apprehended. They held him for a while, looked for drugs, and then released him. The episode occurred ashore. Art had too much to drink and drove through the side of a drive-in restaurant. His excuse was he thought it was the way to go to pick up food. Later on, the Lilly brothers sold the yard to Bert Javin, the tyrant of Back Creek. After two years, we moved the boat over to another yard nearer Annapolis. From there, the boat went several years later to Carolyn and Ron's.

At the time I left Annapolis and came to Baton Rouge, I gave the boat to my daughter Carolyn. The *Escape* is still afloat and Carolyn's husband Ron uses her as his office.

Chapter VII

The first years of operation were in the basement of 1600 Longfellow Street, which was the headquarters of the firm. As the firm grew, we moved from 1600 Longfellow Street to 1317 Vincent Place. Vincent Place was a small town house or square occupied by other firms. Much of the activity discussed in other chapters was conducted at 1317 Vincent Place until we changed over and began to do larger work. Work for the US Department of Energy was conducted at 1317 Vincent Place. It was my intent to buy 1317 Vincent Place and the town house immediately adjoining it. However, the firm could not complete the deal so we looked for another space. We moved into Elm Street across the way and stayed there for four years.

Following the move to Elm Street, we became engaged in doing some asbestos work. This was the beginning of the end of our existence as a separate operation. We subcontracted with Dewberry and Davis of Virginia to conduct asbestos surveys in schools and other public buildings in the District of Columbia. The Firm experienced an unprecedented growth. In a matter of weeks, we grew from a small firm to having fifty employees. During this time, we also undertook in the State of Virginia a set of inspections of school buildings for asbestos. Although this work was very interesting and very fruitful as far as the firm was concerned, it was not very economical. I will not go into the causes of our financial crisis, which occurred because of long duration between completion of jobs and payment by the clients, and we had to declare bankruptcy.

After exploring several possible ways of doing business, we filed for bankruptcy to allow ourselves to operate and continue business, while paying off debts straight up. Henry Counts was our bankruptcy lawyer, and helped us a great deal from then on. We made a move from Elm Street to 1364 Beverly Road and began reorganizing our company and operating elements more conducive to our work. It was my decision and mine alone to continue operation and go back to being a think-tank operation supporting the defense establishment. We emerged from our Chapter 11 Bankruptcy in five years. We successfully paid off all our debtors and all our creditors and did a very good job.

During a ten-year period, we returned to our worthwhile approaches of doing business. From the highly laborious, intensive work of the asbestos, we returned to more satisfactory work of an intellectual nature. The asbestos work was highly labor-motivated and required much on-site work. A large number of approximately fifty of our technicians were reassigned to all the companies doing summer work.

Earlier, while we had engaged in work with the Coast Guard, at the same time we were doing asbestos work. We now focused primarily on the former rather than the latter.

In the early 1990s, we began a full-scale occupational health and environmental engineering firm. Early in 1994, we began planning to compete with the Navy on a contract to support naval supply systems command. This five years of effort was one of the most important in the history of the firm. It encompasses a large number of reports and activities, described in Appendices I and II. Among the outstanding items were the following:

(a) Substitution Manual – A Navy manual on methods prioritizing substitution of hazardous materials. It involved the requirements of existing directives and a new approach to them.

(b) Participation in Navy exhibits with the Armed Forces, with the National Defense Establishment at San Antonio, Texas.

(c) Work on the Navy Regional Concept for Support of the Fleet.

I have made numerous contributions and engineering designs in my career. Following are some examples and not necessarily presented in chronological order.

Asbestos Removal from the Dome of the Library of the Navy Astronomical Center at the Naval Observatory

This project involved studying the contamination of the dome of the library and then designing a solution. We first made some tests to verify the presence of asbestos. These were positive. Asbestos in the dome was flaking off and contaminating the books. The desire of the chief of naval operations was to keep the library open and not close it. This complicated the issue.

We finally decided to isolate the dome by building a ramp with a platform on it, then decided to remove the asbestos and replace it with a non-asbestos mixture. During this work, we were able to locate the original drawings for designing the facility. These we used in the final engineering designs. The project also involved maintaining for the government a series of observations during the removal. All were satisfactory. Studies were made on the books, which were very valuable books. There was no indication of an unseemly amount of asbestos accumulation.

Visibility Study in the New Design for the White Oak Paint Shop

The existing paint shop at White Oak, Maryland, consisted of a waterfall-type paint shop. We made a number of observations on the existing facility and did a feasibility study. This study engaged several options. One was to contract out the work. Another was to build a new waterfall-type spray booth. A third one was to purchase a new facility and install it. This was my first experience of an in-depth study involving the Navy Manual T442 behind the economic analysis. All my previous work at VMI involving economics came into play. This was a classic make or rent facility.

After a long time and many studies, we decided that the most economical thing for a long-term operation of the facility was to install a new dry-type filter system commercially available. This facility became a prototype for many others that we designed in the course of our observations, correcting occupational, safety and health deficiencies.

Correction of Occupational Safety and Health Deficiencies Navy-Wide

This project involved visiting some eighty Navy activities. Earlier, we had conceived a new type of cost estimate, which was based upon the Sweets Manual. We computerized the system and were able to use it successfully throughout the four-year exercise. Involved were corrections of occupational, safety and health deficiencies in Navy shipyards, Naval facilities, Naval air test ranges and so on. It involved correction of battery charging operations, paint spray booths, and many other types of facilities. We used the Manual of Industrial Ventilation prepared by the American Conference of Government Industrial Hygienists, an economic analysis model for computer-driven systems and otherwise the good judgment of a designer. Several hundred projects were developed. We provided drawings of the facilities, the completed DV Form 1191, and the cost estimates. The results were written up in addition to this and were published, as the projects were approved by the Navy Energy and Environmental Support Activity. Many of these projects were actually built and were within the cost estimates by ten percent.

The only projects which were not approved by us were correction of steps in the Naval shipyard in Philadelphia, and the hospital area of the same station. It was our judgment on each case which was the guiding principle.

The Design of the Navy's Hazardous Substitution Manual

This project took several years. It involved a lengthy study of various regulations and requirements for different methods of using economic analysis. The final result was a blend of existing directives (Department of Defense in Navy, Air Force and Army), which allows the user to select the least harmful alternatives. This was published in the book *Environmental Engineering Encyclopedia*. A dozen approaches to the task of selection of least harmful alternatives, cost-effectiveness of solution, and direct cycle of the installation of the facility was suggested. In approximately three years, it was published in the Navy system. An example of this system's new algorithm is shown in Appendix III.

The Effects of Noise on Whales

An interesting project was the analysis of the effects of noise from sonar on whales. With the lack of understanding of the basic principles of noise, it was assumed that there was an aftermath of noise on sea life, several hundred miles from the source of the sonar. An in-depth study was conducted by us that showed that there was no effect of the para levels in question. There is a near-field effect that causes several problems affecting humans in the water near the sonar. The Navy has studied this problem extensively, and the facts are that no damage can occur in the significant distances.

Navy Manual for the Navy Requirements of Hazardous Materials Control and Management

We conducted a series of studies throughout the Navy on hazardous materials uses, control and management. It was recognized that the Navy had to use certain hazardous materials. A method of control and management was devised. It involved selection of the use of material, following the application of a previously described system and the judgment of the commanding officer. An authorized use list was devised for application aboard ship and installation. Its implementation varies in the Navy, at the present time.

Coast Guard's Hazardous Materials Usage

I was retained by the Coast Guard to look at the problems aboard their ships. A study was made of the various vessels and their usage of hazardous materials. The ultimate aim was a creation of new Coast Guard regulations. Such was proposed, but nothing happened.

Engineering Studies for the Air Force

We provided studies for the US Air Forces Air Combat Command. These studies were involved on various levels. They included the examination of waste water treatment facilities at Barksdale Air Force Base and the water distribution system at Wrangley Air Force Base, among others.

The foregoing are examples of a few of the engineering studies we conducted. They extended over a long period of time, involving a great number of individuals. Later on in the history of the firm, we made other subjects. These included a safety analysis and review system for the fossil energy program for the US Department of Energy, an occupational safety and health program at engineering technology centers, and later on, the commercial projects of various contractors affecting contractors of the Energy Department. It is impossible to discuss all of them in this chapter. In all of them, we applied basic engineering including cost and cost-effectiveness. The underlying thing in modern work is the cost of it. Life-cycle costs are important issues and have to be taken into account. Not only what something costs today but what its future cost and benefits will be are important. It is our hope that the work we have done has been highly valuable for the Navy and to all users of our services.

During the last half of the 1995 period, two related episodes occurred. First was the inclusion of the environment, safety and health aspects of regulation for the acquisition systems and the second was the new attack submarine. We were given the job of providing the DOD with information on environment, safety and health aspects of the new Acquisition Regulation 5000-2R. As a result, we visited the Air Force's Center at Wright Patterson Air Force Base and learned all there was about this new system, after which, we wrote relevant portions of 5000-2R.

This effort was extensive in time. The Regulation involved a wide variety of aspects of the new acquisition requirements. It involved new efforts of two arrangements. The first of these was the requirement and the second was how to achieve it. The regulation of projects was changed to fully meet all environmental, safety and health requirements. We successfully wrote the sections pertaining to new systems and the existing ones. Also, we wrote the new Navy requirements.

Following this, we were given the assignment of the new attack submarine. That effort lasted for two years and involved a wide variety of operations. One of these was the use of existing equipment off the older submarines, as well as new requirements pertaining to selection of hazardous materials, and the documentation of where all the hazardous materials were used.

At the outset of the program, we initiated a system of documenting and keeping available all information on hazardous materials to be used on the system. We reduced the number of materials used by combining existing requirements into single documentation. We worked with the Electric Boat Company in developing a system of collating a list of prime sub-contractors as well as the work with the original boat company. A training program for all classes of employees was also initiated. At the end of the second year, the Electric Boat Company and the new attack submarine program received the Secretary of Defense's Award for outstanding efforts. Following that, the next year it also received the award again.

This effort received widespread recognition. The new approaches are now firmly embedded in the Navy systems.

Air Force Regulations and Procedures on System Safety

We worked on projects for the Air Force for a number of years while at the Surgeon General's Office. The directive 5000-2R included the basic principles of system safety. We used the same principles of engineering for the Air Force as we used for other departments of Defense. These were: it was easier to correct the deficiency area in a particular operation than later on.

Also, environment, safety and health are independent with related operations. In each of the following, we have endeavored to pursue basic principles as outlined in the DOD directive and others preceding it.

Analysis of a Proposed Twin-Screw Extruder and Explosive

From a Navy Ordnance Lab, a preliminary hazardous analysis was made of the proposed adoption of a commercial twin-screw operation. This extruder was made in Germany and used in commercial pill production. A detailed analysis was made of the operational system and the instructions. This project was complicated by the fact that everything about the system was written in German. We accomplished the hazardous analysis and made recommendations as to corrective measures to make it safe.

Analysis of the Improvements to the OTTO Fuel

A comprehensive review was made as part of the survey for the rehabilitation of OTTO fuel production facilities at Indian Head, Maryland. A comprehensive review was made of the various stages of production of fuel and associated items. It included on-site sampling and other measures for workers' protection.

Last-Cycle Support of the Navy's Deep Water New Attack Submarine Program

This effort involved large-scale operations of the Navy's deep water undersea operations. A comprehensive report was prepared.

Work on the Contract of the Fossil Energy Program of the Department of Energy

Assessment was made of the life cycle impact on the fossil energy program. This led to the publication of a fossil energy program regulation dealing with the environment, safety and health.

All these related efforts had a profound influence on subsequent developments. In a consultation program with the Navy, we utilized the same principles as we had learned in work for the Air Force and the Department of Energy. It is noteworthy that regardless of who is in charge, the principles are the same. No matter what the program is called or who runs it, the principles are those which we had outlined.

It is noteworthy that all these efforts were recognized when I retired and I was awarded the Distinguished Service Medal (see Appendix IV). It is the highest award and is evidence of the importance given to my efforts for the Navy and elsewhere.

Chapter VIII

This chapter will provide some pictures of the various houses I have lived in. The main chapters detail my life and times and that of others throughout the thirty years of my military career and thirty years following. This chapter will give an insight of the houses we lived in. 519 Fannin Street, Shreveport, Louisiana, was my earliest home that I remember. Fannin Street was the original wealthy street, so was Travis Street. However, the Jones family lived at 400 block of Fannin with Molly Ellerbe living in the Ellerbes' house across the street from 519 Fannin Street. 519 Fannin Street was an elegant old house. It had been added to from time to time by my grandfather. The main house was separated from the servants' quarters by a yard containing a fig tree and other items. There was also a large garage in the backyard. During most of my youth, the servants' quarters were occupied by my maid and the household chief of staff Mrs. Ella Fitzgerald and her husband James.

My grandfather and grandmother had a suite at the top of the house, which was a large bedroom. I remember having occasions to gather there in the evenings to listen to the radio. My mother occupied a bedroom across the hall from my Uncle Albert Henry. Later, I had that room. My Uncle Milton occupied the front bedroom for a while after his marriage. Then, it was rented out. Across the street were the Ellerbes and the Dolls.

Shreveport, during the time I was growing up, experienced a boom because of the growth of the Homer and Haynesville oil fields. In spite of the Depression, which started in 1929, it was a lively place.

After I was ordered to active duty in 1941, I had received 532 Forest Avenue as part of the settlement of my grandmother Rosa's estate. 532 Forest Avenue was of Spanish architecture looking very out of place between traditional Southern homes. However, it was a lovely home. My grandfather Abe and his wife Rosa had occupied the house after he left his home at 1400 Louisiana Avenue. That house was still in the estate. It was occupied for years by Peachy Gilmer. The Gilmer Chest Clinic was the principal resident.

When I was stationed at Fort Bragg and then at Fort Jackson, I visited home quite often. At Fort Jackson, where I got married, I was stationed first at the Louisiana maneuver area and then at Fort Leavenworth. Viv and I occupied a rental apartment there. The next several months after a Command General Staff School, the desert training area and then at Camp Roberts, California, I lived in a motel. From there, we drove across country to Carlyle Barracks where we were stationed to take a course at the Medical Service School. For the several months we lived in Carlyle, I was living in a lovely little apartment above a restaurant and bar at Mount Holly Springs. When school was over, I left, going to Shreveport to stay at 532 Forest Avenue and then on by way of the mountains to Salt Lake City. There we finally bought a small house in the Sugar House area. That house was a lovely, small home.

My daughter Carolyn was born at the hospital in Salt Lake City. From there, I got orders to go overseas. I took my children to my in-laws in South Carolina, who kept them while I went on to the overseas replacement depot. From there, we moved on to the overseas area at Bremerhaven and then into First and Felbruck. Events there are mentioned in the main text. We lived at 28 Rhineblick Strasse in Wiesbaden, Germany, for three and a half years, which was quite an experience. We became very well adapted to life in Germany. We left Wiesbaden for Dayton, Ohio, and flew on American Overseas Airlines. I arrived in New York, and then went on to Boston to see my Aunt Minnie and Stuart Newland, her husband. I bought a car and drove to South Carolina and then to Louisiana. In Louisiana, we stayed at 532 Forest Avenue, and then moved on to Dayton, Ohio. In Dayton, I rented a house and looked for permanent quarters near the base. We bought a house at 16 South Street. The small house was a lovely place because, after two years, we installed a second addition to it in the basement, a full-length underground basement bedroom in which we lived. It was a lovely place and we had a great time.

At Wright Patterson, we drove to Omaha. It was the middle of summer and it was quite an undertaking. I took the Cadillac convertible and Viv drove the station wagon. I took Felix with me and Carolyn went with her mother. We stayed at the Hotel Blackstone when I first arrived in Omaha. The Blackstone was an ancient old hotel but very well equipped. We found a house near the base, near St. Andrew's Church, and near the Aksarben Race Track. We rented it for a year and then found another place to build a house. 1502 South 94th Street became our home for the next six years. We joined the Omaha Athletic Club and enjoyed its festivities very much. Then, I received orders to go to the Industrial College and was supposed to come back to SAC, so I kept the house. The only memorable thing was that I found in the basement of 1502 some barrels which had not been unpacked. I said, "Let's get rid of them. We haven't missed them or know what is in them, so let's throw them away." Vivian objected saying, "It might be family heirlooms." I said, "Well, we have had them for six years and may as well get rid of them." So, we did.

We rented a house near Lake Barcroft. After I got my orders changed and had to stay in the Surgeon General's Office, I found a house being built in McLean, Virginia. We bought it, and there we stayed for thirty years. 1600 Longfellow Street was a great place. It caused me great grief to sell it and leave it, but at the same time, we moved to Baton Rouge where Grace had a house already. It was a house that my Uncle Milton had built and lived in since 1940 or so. It is where I now have a room as an office and am writing this book.

These are the fine places I lived in and certainly enjoyed. My wives and I have had a good history of living together and have had many fine hours. Each house became a home. Homes can be simple or elaborate, but the central element is the people who live there and their attitudes. I hope that mine were good houses that became homes.

Someone not mentioned elsewhere in this book is Sandra T. Wisneski, the Vice-President in the early years of the Meyer Corporation. She had served in the Nixon White House, and left because of her pregnancy. Her services have been valuable. She later divorced her husband and moved to Asheville, North Carolina.

She was replaced later by Joe Hummerickhouse. Joe had served with me at SAC as a junior officer. He later left the McLean office to retire to Florida. The final organization of the Corporation is shown in Figure 1. Several staff officers are shown.

Chapter IX

It seems fitting to make comments about some of the men I have known. During the years in my professional practice, I have met a great number of men who have had a profound influence, in one way or another, on my life. Herewith are some of my recollections about these people. They are not arranged in accordance with any time schedule nor did I try to put them likewise. First and foremost, of course, is Carter Haines, Professor of Sanitary Engineering at VMI. Carter was a great man who later on went to serve in the Inter-American Affairs Commission with Rockefeller. Thanks to him, all of my energies at VMI were directed towards more and more learning about engineering and carrying out my future. Another truly great man was Abel Wolman. Professor Wolman, head of Sanitary Engineering at Johns Hopkins University, had more to do with my professional career than almost any other person. I never took the opportunity to attend his courses, but I was deeply indebted to him for his personal interest and his leadership. Later on in my professional life, I had an occasion to have lunch with him at the Johns Hopkins Faculty Club. He was ninety-eight years old at the time and just as vibrant as ever. When he died at a hundred plus years, a truly great man expired.

Otis Schreuder was another great person. Otis was a Colonel in the Medical Corps when I first met him at USAFE. He later became a Brigadier General. His sterling leadership led me further down the road. Dr. Schreuder was a man ahead of his time, a great leader and a great person. At SAC, Lloyd Griffis was another fine individual, a great leader and a sterling character. Griffis was a man far ahead of his time also. He taught me a great deal, to be a military officer first and then a medical service officer later. The six years I served at SAC gave me ample opportunity to meet a number of individuals, who later I had a nice time with at the Pentagon. Among these were General Blanchard, General Jack Ryan, who was Director of Material and later became Vice-Chief of Staff and Chief of Staff of USAF; and General T. S. Power. These men made up a bureau of military intellectuals. I felt at ease among them and relished their friendship.

No mention of my SAC experiences would be complete without mentioning the three membered Air Force surgeons, Colonel Patient, 2nd Air Force; Colonel M. Towner, who later on became Surgeon Generals, and Colonel Bohannon, Surgeon of 15th Air Force. I knew them well in the past, as well as in the present and the future.

Among those I knew in the biomedical science professions were: Robert Peterson, Industrial Hygiene Engineer; Jack McCambridge, a learned individual who later left BSC to become a project officer; Clarence Feightner; and a host of young engineers and some not so young appear in the annals of BSC. Herbert E. Bell, who I recruited at VMI, was another outstanding leader who deserves mentioning in these pages. Also worthy of recognition is Ed Poth, another industrial hygienist at Kelly Air Force Base. He laid the pathway for many others to follow. I am not through yet with documenting everyone. My apologies go to them who are not mentioned because of time and other constraints.

I have neglected many people I have met in the last twenty years. However, there are many fine individuals whom I have met and whom I respect. Among these are: Dave Price, Director of Industrial Hygiene and Pollution Prevention, John Hannum in that same office, and Dr. John Talbot, Brigadier General, USAFMC, who served in the Surgeon General's Office with me and later on was a close friend, as mentioned earlier.

The names I have listed are those of immediate friendship and knowledge. Others I have known and worked with are a part of a respected past. I have mentioned them elsewhere in this book or elsewhere.

Chapter X

MISCELLANEOUS

An Unexpected Find

In early 2000, I received a letter from a company in Texas stating that they were putting together a new lease on 804 acres owned by Felix Halff, my great-grandfather. I am the oldest grandson of Abe Meyer. His wife Rosa was Felix Halff's daughter. My middle name Felix is a carryover from Felix Halff.

Mr. Halff apparently left Texas and moved to California. In his early days in Louisiana, he was quite a character, apparently something of a devil. According to rumors and legends, he secreted a large quantity of cotton in New Orleans which was seized by Union forces. Later, after some years, Congress voted him a substantial sum of money in return for the Feds having seized the cotton. While writing this, I am undecided as to what to do if this lease becomes a viable financial interest. I am considering making Felix and his sons principal heirs after me. We will see what happens as a result of this substantial lease. The surface has on it a nature park owned by the State. We own only the mineral rights.

I have asked Roy Beard to look into other areas that may be involved with the Halff Estate. The lease is for 3/16 of a total of oil and gas on the 804 acres. The wells may or may not come in; therefore, we do not know if this lease is available or not. Assuming it is, we will have to take a long look at it and decide what to do later. In the meanwhile, it is a valuable and interesting item in the history of the family.

Random Thoughts

JOE HUMMERICKHOUSE

One of the first to come to mind is the relationship with my Vice-President, Joe Hummerickhouse. I had known Joe before, while I was at SAC. He was a Lieutenant and a Captain. I followed his career in the Air Force, and when he retired, he was a Colonel at Air Force Systems Command Headquarters at Andrews Air Force Base. We were invited to his retirement party.

I questioned Joe as to his plans when he retired. He stated that he had none. I decided to offer him a job with A. F. Meyer & Associates, and he accepted.

His first job was as Director of Operations. He readily took over and became a very important member of our staff. He aided and assisted me in many ways, later becoming Vice-President of the Company, and then he moved to Florida.

Among his many accomplishments was serving as the Company's representative to the various working groups of the Navy's Pollution Prevention and Environmental Affairs Committees. Joe's abilities were varied and manifold. We will see more of him as we go through several chapters.

JACK MCCAMBRIDGE

Jack was a very good friend of mine over the years. While he did not have anything to do with the Company, his friendship has been of considerable importance. I have mentioned him briefly earlier. One of the outstanding memories I have is his annual Christmas parties. These events were very much enjoyed. I remember one at which General Tanner at his retirement was present. Tanner in his retirement had shown advanced age. I did not recognize him at first. Jack had built a whole new house, a very exclusive and fine residence.

RON SILVA

Silva was an excellent officer. He retired and lived in Albuquerque. We visited him during the balloon events in 1998. Ron gave us a great time. The event was attended by over a hundred BEEs (Bio-Environmental Engineers). It was a far cry from the five or six on active duty at the beginning of a BEE program.

LOUISE MCALLISTER

This book would not be complete without reference to Louise McAllister and the Office of the Environment, Safety, and Health for the Assistant Secretary of our Research Development and Acquisition. Until she retired, she was a vigorous supporter of this office. It was through her efforts that we undertook the work on the Defense Acquisition Regulation 5000-2R. She was not only a vigorous supporter of our work, but a good personal friend as well. Her retirement was sorely felt because of her wisdom and abilities as a manager.

CARS

Throughout my life, there has been a fascination with the automobile. Like many other Americans, I grew up wanting to own a car. In 1941 when I graduated, I bought a Chevrolet. This car was a two-tone coupe with white sidewall tires. After I married Vivian in 1942, we drove the car until the tires were worn down, about 20,000 miles. Then, we traded the car for another Chevrolet. After that, if my memory is correct, we bought a 1941 Buick. When I left to go overseas, I sold the Buick. Vivian had her own car by then, another Buick. This we kept until she came overseas, after which we bought a new French car, a Renault. This tiny car was great and I drove it from Paris to Wiesbaden and all over Germany. We traded it for a 1945 Chevrolet when one became available. This car was subject to my making a new transmission for it when it failed.

When I returned to the States in 1945, I bought a Packard. I remember the guy buying it back from me and being told that he could write a check for as much as he wanted. At Wright Patterson, I traded the Packard to my assistant and bought a new Cadillac. From then on for several years, I got a year-old Cadillac from the General Manager of their division in Dayton. In addition to the Cadillac, I bought a Chevrolet station wagon with a Henry J. and subsequently another Chevrolet for Vivian. While we were in Omaha, I bought new Cadillacs from the dealer. Also, I traded the Chevrolet for other cars.

I became irritated with Cadillacs when I got one that leaked water into the doors, so I traded the Cadillacs for a Borgward coupe. This German car served me well for three years. I drove it from Omaha to Washington. Later, I traded it for a Ford station wagon. I traded the cars about every two years. I bought an Oldsmobile convertible and an Oldsmobile station wagon. The station wagon lasted for 125,000 miles when I gave it to Carolyn and Ron. What I am currently driving is a Lincoln Town Car. It is an excellent automobile with very good driving abilities. All in all, over the years I found General Motors cars to be very good and equal to anything on the road.

MY HERO

While at Wright Patterson, I had a great time with a horse named Hero. Molly Schreuder owned him, and Jimmy Dale and I rode the horse quite often. He was a spirited animal.

I really remember one night. Molly and Otis said over the phone, "Hero has gotten loose over at the stables. Will you come over and help us get him up?" Well, I was getting ready to go to the Club and told Vivian to wait for me and I would be back in a little while. Hero had gotten out before and it had been no problem at all. This was a different episode, however. I left the house at 16 South Street and rushed over to the stable in which we kept the horse. Well, he had kicked out the stall and was running loose around the field. Murray and Otis were there, as was Jimmy. It became evident to me that we had a real problem because the stallion Hero was reacting as stallions did. It was going to be a long night, I thought, and it was because the horse was in a lather. After about an hour, I said, "Give me a halter and I will try to get him." I took the halter and went down the field and tried to catch him again. He was in no mood to be disturbed by anybody. After an hour or more, I got enough courage to approach the horse gently and captured him. It was a wild night.

Hero was one of the original Austrian stallions. He had been liberated by a friend of Molly's in operations in Europe. I had moved him from Germany to my father-in-law's plantation in South Carolina. He was then moved up to Wright Patterson. Quite a horse! Vivian and I had many fine times with Hero. He was a favorite of the horsemen around Wright Patterson. After Otis retired, she moved Hero up to Connecticut.

BROWNSVILLE, TEXAS

During the cold winters of Nebraska, I became involved with life at Brownsville, Texas. My friends, the Freemans, and ourselves became interested in moving south for the winter. I announced it one year that the time had come to take a vacation in the South and thought it would be wise to go as far as we could and stay in the United States. Jane Freeman agreed and told her husband Max that we should go somewhere like that. With snow on the ground, it seemed like a good idea to be going south.

In February, we decided to go. I went down to the School of Aviation Medicine at San Antonio to give a speech. From there, I went on to Brownsville to be joined by the girls, Vivian, Jane and Max. Brownsville at the time was a sleepy, southern Texas town, the farthest south within the United States.

We stayed at a motel, the only large one in town. This motel became our headquarters for several years thereafter.

It was really great to get away from SAC Headquarters and the snow. While I loved my work, I enjoyed the sunny Texas climate for the winter. We traveled over the South Texas area quite a bit. Joe Hummerickhouse developed a set of plans for the land we had bought.

All in all, Brownsville was a lovely place at that time of the year. We played golf at Padre Island and had a magnificent time. In retrospect, this was one of the calmer periods of the last few years. There was no time to worry about the world and anything else but just have fun.

ASBESTOS IN SCHOOLS

I had been engaged for a number of years in work on asbestos in schools. We lost our shirt. I said to myself several times, this is a losing battle. However, we continued it because it was necessary.

We had been involved with Dewberry and Davis, in the District of Columbia, doing asbestos work in public buildings. Following that, we engaged in competing for work in the State of Virginia. It became very apparent to me that the school work was not the sort of work we ought to be doing. However, we had to keep up with it. The principal difficulty and the difference between this work and the standard Virginia work was that instead of paying at the time the work completed, they paid after the complete work was reviewed by the State. Inevitably, there was a long period of time between completion and review.

Finally, I appealed to my friend J. C. Wheat, who was a prominent investment banker in Richmond. He went to the State and asked them to speed up the process. Very little resulted. It was not until we went down to the Capitol and visited the State offices that we had any results.

In a number of cases, the choices were very simple. Either remove the asbestos or apply placard for the future and not disturb it. Sometimes, disturbing the asbestos created a greater problem that was overcome by simply placarding. We took this approach on asbestos in areas that were not likely to be inhabited or occupied.

In addition to removal, we had a contract to train the personnel involved in removal. This was very lucrative and we got paid more promptly.

WRIGHT PATTERSON AIR FORCE BASE

Dodge Memorial Gym was a great feature of life at Wright Patterson Air Force Base. This magnificent facility, which was adjacent to the officers' club, was built by the Dodge family. It contained an Olympic-type swimming pool and badminton courts, as well as other facilities.

Vivian and I used the pool and the courts nearly every evening during the winters. The indoor pool was a lively adjunct to the outdoor pool at the officers' club.

I played badminton quite a bit. It was a lively sport and I enjoyed it very much.

The officers' club pool was a very good facility also. I gave my kids challenges to swim across the pool and then the length of it. There were lively contests between the two of them. Carolyn was the first to swim across the width and was then followed shortly after by Felix, who was older. Dodge offered me a great deal of relief at a time of working hard and playing hard. We got to know the club officer Major Knox Booth quite well. He took us on several trips. I had met him when he was a club officer overseas in the Philippines. These were lively times and much enjoyed. The house at 16 South Street became a regular place for the VMI crowd to meet on Fridays and Saturdays. My friend and colleague Jimmy Dale was a regular visitor. Jimmy was engaged at the time and on the verge of becoming an ATLO (Aero Technical Liaison Officer). He attended the school at Wright Patterson and was a very, very bright young man. We have met each other many times afterwards, and he is one of my closest friends.

We cannot mention Wright Patterson without talking about life in the officers' quarters. In addition to the brick quarters, the wooden structures, which were occupied by Colonels and Lieutenant Colonels, were behind the headquarters building. Among those whom we knew were Jim Humphreys and his wife and Colonel MacEachern and his. I mentioned MacEachern before. It was an easy-going relationship between us. Colonel MacEachern was an avid painter. Although MacEachern and I were professionally sometimes at odds with each other, socially we got along really fine.

All in all, life at Wright Patterson was full of fun and games, as well as hard work. It was a very fulfilling experience.

TALES OF A DOG

Throughout his twenty years or more of life, the dog Rebel was the main feature of our life. Rebel was an English cocker spaniel given to us by Mrs. Schreuder. As a puppy, Rebel was an outstanding rebel. It was not until he had eaten up the cuffs off Vivian's coat that I decided he had to go to obedience school. After months of training, he was an ideal dog.

In addition to the exploits mentioned earlier, there are a number of other things about Rebel which need recording. He became a boat dog early on. We would get ready to go to the boat and he would be in the car first. Frequently, when we arrived at the boatyard, he would jump out and go down the dock ahead of us. All in all, he was a very good boat dog. He cruised with us over the length of the bay and was a fine, fine companion.

When Rebel died, a void existed for a long time in our lives. Vivian was very much depressed by his departure. So was I. A long time thereafter we had no pets, but a cat across the street from Mrs. Fitch's came into our life later. I rescued him from deep snow one evening. We had had a snowstorm that left over two feet of snow. The cat was perched on a fence between our house and the house next door and jumped into the snow. I rescued him and he came to live with us for a while. Later, much later, after Vivian's death, we acquired another dog. He was a German Wiemaraner and a lovely dog. He would not go to the boat, however. Later, there were other dogs and cats that replaced Rebel.

CRUISE TO EUROPE

In the summer of 1998, we decided to go to Europe on the QE2. Grace and I left Washington and went to New York. We stayed at an old hotel around Central Park. The hotel was being remodeled and was not up to our expectations. However, we called Henry Jacobs and he and his wife Charlene met us for dinner that evening. We discussed with Henry the problems we were having in Louisiana with the trees. We decided to put our shares to work. Henry met us the next day when we were boarding the QE2. He had met with the purser and arranged for us to attend a number of parties. At our table on the QE2, there were a couple of ladies from England. They had taken a ship in Southampton and come to New York and stayed aboard till the time we went back, a fascinating couple. We also met a senior Concorde captain and a foreign assignee of the State Department, plus an FBI agent who was going on an assignment to England.

The highlight of the trip was the parties. In addition, we met a British woman who was a RAF widow. We landed in Southampton and took a car from there to London. Overall, it was a lovely trip. Grace was thoroughly fascinated with the English landscape.

THE SST AND OTHER THINGS

Later on during my stay at EPA, we became involved with Concorde. The EPA's position was that the plane made too much noise and was unacceptable. I was sent by the Administrator to England to discuss with the British and later with the French the facts of the situation. There was great fear in the EPA that hundreds of these planes would be built. Accordingly, I was selected to go and find out what the facts were.

British Airways had a Concorde scheduleed to fly into Dulles Airport in Washington. I booked two seats on it and took off for London with Vivian. We arrived after three hours and some minutes. We went by the British facility associated with the airplane itself (British Aerospace). They were very open and told me everything I wanted to know. We visited the aircraft factory near Bristol. There were only a few planes being built. I asked them how many they planned to build and was told twelve or thirteen, a far cry from the hundreds thought of by the EPA advocates. We had a lovely time in London. One night, we traveled to the mouth of the Thames to have dinner at an inn there, a stone building built in 1400. It had seen many individuals, and had a long history. The owner was an RAF widow. She was very lovely and showed us everything we wanted to see. Vivian went on to see Felix and I flew back on Concorde. On my return, our report indicated that there were only twelve airplanes being built by the British and that the French accordingly were willing to buy about five to seven of them. I reported that the airplanes were noisy but were not very much so, and that we should accept them on the basis that they would fly only from New York or Los Angeles and that they were acceptable otherwise. After a long delay, this was the United States's proposition. A number of years later, the airplane flew into Washington and was picked up by Braniff and flown subsonic to Dallas. This arrangement ceased when Braniff was sold to American Airlines.

AFTER SST

Upon completion of the time with British Aerospace, I flew down with Vivian to Italy to see Felix. We flew on British Airways down to Venice. Vivian and I were happy to see Felix. After the usual fuss over the kids and the grandchildren, we settled in for a lovely stay. We decided to go on a motor tour. We visited the Leaning Tower of Pisa and several other famous spots.

It was a beautiful time of the year. We took the gondola tours in Venice and also walked all over the city, on the banks of the streets. We visited the cathedral and the other places that are well known.

Later on this tour, we visited Florence and stayed in a lovely old villa. It had modern accommodations and was very nice. While in Florence, we visited the castle on the hill, across the river from the main city. We also visited the art gallery and the statuary. Florence is an interesting city, one of the finest in Europe. We visited the original statue of David and several other museum pieces.

On our way back, we went to a fine restaurant up in the mountains. It was a typical Northern Italy establishment. We had a fine time and great eating. Later on in the evening, we were serenaded by the staff.

On our return to Felix's quarters, we separated and I flew back to London and then to the States. That was an excellent trip and a very fine one indeed.

On the SST flight home, I was able to be in the cockpit for two hours. This was a great event and one I enjoyed very much. It gave me quite a perspective on Concorde at that time.

Some years later, while aboard the QE2, I met the senior captain of Concorde, as just described. We had a fine time talking about our experiences.

VMI

I will attempt to give a full picture of VMI, as seen with the eyes of an alumnus. Throughout the years of my high school and up to the time I matriculated, the one episode that I always wanted to go through was to be a cadet and graduate. During the war, I had very little contact with the Institute as such. However, I ran into a number of graduates who were either contemporaries or juniors. Among these were those previously mentioned as well as others. For example, while in Germany I met Colonel Felix Feamester, who was a surgeon of the second brigade of the constabulary located in the wine factory in Wiesbaden. Felix was in the class of 1937. We became close friends.

In addition, I met General Burres, the Inspector General of Vienna (mentioned in Chapter IV). During WWII and my tour of duty in Germany, I came to the States and ran into a number of VMI graduates throughout these years. In addition, I became very involved with the Civil Engineering Department. I founded two awards in 1956 and continued them in the following years. The VMI Sanitary Engineering Award became the Alvin F. Meyer Environmental Engineering Award, one to the senior and one to the second classman who led the class in Civil Engineering. Many of these cadets later were leaders in the field.

I could write a whole new chapter or a whole new book on my experiences at VMI. Among the outstanding things were: the dedication to General Marshall of martial arts; football games; the VMI event for graduates who participated in WWII; and many others.

Outstanding among people whom I have met over the years is Jimmy Dale. Beginning in 1941, going from Wright Patterson to date, he is one of my closest friends. Jimmy is a fine officer, and since his retirement lives in Wiesbaden, Germany. He is a frequent correspondent and a lovable individual. Another individual who I have met over the years is Jim Morgan, former Dean of Faculty of the class of 1942.

EPILOGUE

This book has been written primarily for the benefit of my children and for the benefit of my family and friends. Some time later, I am going to write a book for the general public. This one, however, stands alone.

It is said that behind every man, there stands a woman. Well, there have been three in my case. I have to take my hat off to all three of them for having put up with me for all these fifty-five years. Vivian for all her problems was a good wife and a good Air Force wife and otherwise contributor. Sue in her way was good for me as successor and Grace is outstanding, No. 1 and the best.

I now find that one of the most important things to me is my home, where I have many fine years to write a new book or add new chapters to this one. I have not decided that yet. In the meanwhile, each person who reads it will find, I hope, a spark for them to write their own book about life, about events and about the future. This book has no ending because every day something new is being written. I hope readers find it a marker in their own lives, telling them where they are and where they are going. Through it all, I have had a steadfast belief in God and in the Supreme Being I rest my case. I have lived and fought well. I am convinced I could not have achieved all the things I have done over the past years without the help of the Lord. I have lost some fights and won some. Those I lost, I have learned a great deal from; those I have won, also. I hope that in the coming years, I'll have as much success as I have merited.

Someone asked me, "What is the most important thing you have done in all the years of your life?" I would say that there are two things that I have done, which are most memorable and most important. These are the foundation of the American Academy of Environmental Engineers and, secondly, the creation of the

United States Air Force Biomedical Science Corps. These two events, although separated by many years, are of much importance. The first one was the creation of a body of expert civilian and military communities. The long-range step made after much difficulty is of continuing importance. The creation of the Biomedical Science Corps, both in its time and today, is of great importance to the Air Force and the nation. These two events are among many others which hold major importance in my life as well as to the military. I consider them to be my most important contributions to science and to man. There are many others, but these two outweigh them all.

When looking at the contribution of the two, I find them of equal importance and do not consider them most important in order.

The Academy has achieved, on its own behalf, wide recognition in the engineering profession, and the Biomedical Science Corps stands alone as a role model to guide health professionals. There is talk today of many other organizations trying to find a place under the sun. The Academy's recognizing, as it does, professional engineers, who achieve beyond their professional registration, is an outstanding institution. The Biomedical Science Corps continues to be a unique operation. These two organizations are self-growing and self-sustaining. They have both grown into magnificent structures.

I know not what the next five or ten years or even the next year holds for me. I am happy with what I have done and what I am doing now. To me, the achievements are self-evident. What I have done, I would not necessarily do again today. I would leave them in the past as they are. I have made mistakes, but looking back I would not change anything. As my Uncle Lionel said to me, "I've lived a full life and would not change it even though I would not do it the same today as I did in the past." This has been a guiding principle I have had over the years.

In the spring of 1998, I decided to move to Baton Rouge. We decided to go because of the fact that business was falling off, and we had decided to move on back to Grace's house. As part of the move, I had my grandson-in-law and grandson come to the house and take their share of the household effects. The rest of the things were put into boxes for the move.

To make a long story short, in the move the truck caught fire at Moss Point, Mississippi. It totally destroyed everything we owned. We rode to Moss Point the day we were notified that the truck would not be at Baton Rouge. We found the truck and saw the damage. It was so extensive that very little belongings were saved. Through a fortunate circumstance, my portrait that was given to me at the time of retirement was saved as well as my decorations. The rest of the things were completely burned up.

We filed a claim against my homeowner's policy. It took almost a year to be resolved. However, the homeowner's policy did not cover everything that was lost. We are still engaged in the business of trying to recover. A lawsuit has finally been entered against the owner of the van. The outcome of that suit is uncertain at this time.

We had the office arranged in the house at 9060 Meadowood. The year 1999 found me in Baton Rouge, Louisiana. Trying to get over the shock of the losses, and taking care of my eyes was a big problem. I don't know how I managed to keep up my good spirits and with my work.

After the fire, we began to have all sorts of problems with the insurance company, that is, with the Travelers. The initial start was of $5,000 on homeowner's policy and a large number of questions were raised. Suffice it to say, we finally received the full $150,000 on the homeowner's. Furthermore, we have the suit against the owner of the van. After a lengthy proceeding, the judges of Fairfax County have ruled in our favor. At this point, we are awaiting a final settlement of the claim on the whole amount.

More important have been my experiences to date with the various practices of dealing with Medicare, starting with my having to have a follow-up on my eyes. The eye doctor was a very fine physician. I was having difficulty in seeing and so scheduled my right eye to have repair promoted. My left eye, we decided to leave alone because of the double vision I've had since 1950. The operation was a relative success; however, because of some damage to the retina, I still cannot see. To date, as write this, my personal vision is not complete, and I cannot see beyond ten or fifteen feet. That alone is a problem and what happened next is an interesting story. As a matter of routine, I went to a fine orthopedic man. He suggested that I have a follow-up made on the work done. Later, I began a lot of exercises in futility.

Making a long story short, we went to several doctors. The consultant in neurosurgery said, "There is nothing in this record I can see of any cause for any horror." The consultant and his next-in-chief said twice that there was nothing in the line of Parkinson's Disease (which I already knew). That result was finally the same diagnosis that we had at Andrews Air Force, namely, stenosis of the lower spine. I have nothing but praise for the doctor. My criticism is of the system. It called for repeated visits to various doctors, at the end of which I still had confirmation of basic surgery and basic findings at Andrews Air Force Base. Another thing about the Medicare system is that when we went around the doctor, consultant to the optometrist, to the optomologist about my eyes, he did not have full-scale vision equipment. Also, taking a blood test to confirm what this proved in any of the observations, I had to go to another hospital to get the test made. Adding up the charges of everybody concerned, I became aware of the fact that an awful lot of money is spent for administrative purposes.

On the other hand, I found a very fine dentist. He was able to find a way to isolate and correct a tooth problem I had had for several years. Later, he installed a bridge. Instead of finding me another dentist, he did the whole job himself. Later, I had occasion to take care of our yardman with the same dentist. That evening, I checked and found that he had been there and been to his house after taking the tooth out.

All in all, I found an aspect of Medicare that was no good. The fact that both physicians and their specialists prescribed going to a different doctor for every little thing was deeply costly.

The other thing about the State of Louisiana's medical care is the experiences we had with Grace and the advice problems associated with her. On two occasions, we had to use the emergency room at the hospital, Baton Rouge General. The service was excellent and it took too long and took too many tests. There was no medical officer and I found efficiencies were very low and very costly. On the other hand, the tests were very good and capable even at the late hour. It was my observation that considerable costs could be saved for the individual and the State Medicare System by (a) streamlining some of the procedures, (b) keeping the records centralized so that when one fills out the form, it is in the records, not having to repeat them each time.

The overall impression I am trying to give about this experience is that I do not go to a doctor in the Baton Rouge area without having to do so, except in extreme circumstances. Once one goes by the original doctor and just a number. The situation had to be changed.

I was able to complete in 1999 a number of projects. These included: (a) application to the State of Louisiana pending fuel shortages, (b) the DOD problems associated with fuel, and (c) I sent to Senator Kennedy the cost of OSHA violations.

Also, during 1999, I obtained a charter from the State of Louisiana for a center for independent studies, analysis and recommendations on ESH issues. This affords us an opportunity as a non-profit corporation to obtain priority treatment and submissions on federal acquisition regulations. It also provides a means for me to conduct independent studies.

I have not heard during 1999 of any progress on Newport News Shipbuilding. This is due to the fact that the Company was on strike during most of the year and there was an extensive study as to the needs being made by the Corporation.

During this period of time, we also published an article in the *Louisiana Engineer*. It was devoted to the facts of a larger study of Louisiana problems associated with fuel shortages.

In the first quarter of 2000, the fuel shortage situation became apparent. As was forecast for the first quarter, there was no effort on the part of OPEC countries to raise their prices at the same time as in production. The result was $2.00 per gallon for gasoline. It becomes increasingly apparent that we are victims of a vicious circle, which can be tightened on us at any time. We have a report on the effects on defense that was extracted and sent to Naval Operations. At this time, we are awaiting to see if anything at all will happen. It appears that the public is going to accept $1.50 to $2.00 per gallon. It is an interesting time because the United States is going through an economic recovery. It is an election year for the President of the United Stares and we do not expect any action on our proposals. Yet, sometime in the future, the prices will continue to rise and our need for action and vision in our report (see Appendix III) will become apparent.

During the third quarter, I received information that a BEE meeting will be held in New Orleans. The Association of Retired BEEs is having another session. I am expected to attend it and give a seminar of my observations about the future. This I will work out with Warren Hull. Warren is a good friend and an old BEE. Also, I have written Frank Parker about some work on gasification which we had done a number of years ago. He is a retired BEE and will be at the meeting. I am hopeful that this is the last of the meetings in which I will be able to give my views on the past and the future. I expect to continue my own work hour.

Meanwhile, I am going to keep on working. My health is reasonably good, and I hope to find years of interesting work ahead.

Appendix I

DELIVERABLE WORK DONE UNDER NAVSUP SUPPLY SYSTEM

July 9, 1995 — Trip Report: Collection of Background Information on Regional Hazardous Material Management. Puget Sound Naval Shipyard, June 13–15, 1995

August 2, 1995 — Monthly Assessment Status Report for NAVSUP, Implementation of the Navy Pollution Prevention Strategy

August 4, 1995 — AFMA Support to the RCRA 6002 Interagency Working Group Participants from the Navy (OASM (RD&A) and NAVSUP

August 9, 1995 — Pollution Prevention Life Cycle Cost Savings Case Study

August 9, 1995 — Review and Analysis, Implications to HSMS in GAO Report. "Selection of Depot Maintenance Standard System Not Based on Sufficient Analysis."

August 11, 1995 — Trip Report: Data Collection and Information Gathering for Regional Hazardous Material Replenishment Model, Norfolk Naval Base, July 20 and 21, 1995

June 24, 1997 — Draft Navy Guidance Manual for the Hazardous Material Substitution Process

July 1, 1997	Trip Report: Conference Support for NAVSUPSCOM at the 4th Annual DOD/Industry Product Shelf-Life Symposium
July 1, 1997	Trip Report: Conference Support for NAVSUPSCOM at 1997 Navy Pollution Prevention Conference, Arlington, Virginia
July 1, 1997	Technical Report: Revisions to the DON (Department of Navy) Pollution Prevention Milestones
July 9, 1997	The Proposed Particulate Matter Rule and Its Implications for the Navy
July 15, 1997	Technical Report: Revisions to the DON Environmental Program Presentation
July 22, 1997	Poster Session: Using the Hazardous Material Substitution Process to Evaluate and Select a Material Substitution
August 1, 1997	Minutes of the Environmental Acquisition/Procurement Working Group (EA/PWG) Meeting of July 29, 1997
August 15, 1997	Technical Report: Preventing the Introduction of Non-Indigenous Species into Coastal and Inland Waters
August 18, 1997	DON Section (Discretionary) Defense Acquisition Deskbook Input, Part 4.4.3.7. System Safety and Health, DOD 5000-2R and SECNAVINST 5000-2B, Para 4.3.7.3
August 26, 1997	Conference Support for NAVSUP SYSCOM at the Joint Service Pollution Prevention Conference and Exhibition, San Antonio, Texas

September 23, 1997	Minutes of the Environmental Acquisition/Procurement Working Group (EA/PWG) Meeting of September 16, 1997
October 2, 1997	"RAGMAN" Poster/Slides
October 24, 1997	Minutes of the Environmental Acquisition/Procurement Working Group (EA/PWG) Meeting of October 21, 1997
October 31, 1997	Revisions to the DON Pollution Prevention Milestones
November 4, 1997	ASN (I&E) P2 Milestones with Completed Milestones in Bold Type
November 10, 1997	Acquisition Process Environment Responsibilities
November 14, 1997	Acquisition Environmental Policy Brochure
November 18, 1997	Minimizing Environmental Impacts of Fleet Operations
December 1, 1997	"RAGMAN" Handouts
December 12, 1997	Acquisition Environmental Policy (Legacy) Brochure
December 24, 1997	ASN (RD&A) Acquisition Environmental Policy Desk Reference
January 20, 1998	Deliverable Item No.: JJMA/38/T2/#05 (a), RHMMS Manual, Chapter 1 (Updated); Deliverable Item No.: JJMA/38MA/38/T2/#05 (b), RHMMS Manual, Chapter 2 (Updated)

January 30, 1998	Meeting Minutes for NAVSUP/HSMS Meeting, January 7 and 8, 1998; Meeting Minutes for NAVSUP/AFMA Meeting, January 16, 1998
February 11, 1998	Minutes of the Environmental Acquisition/Procurement Working Group (EA/PWG) Meeting of January 27, 1998
February 13, 1998	Impact of New Particulate Matter and Ozone Regulations on the Navy
February 13, 1998	RHMMS Manual, Chapter 3 TABS 1–7 (Updated)
February 25, 1998	Cost-To-Complete Index Slides
February 26, 1998	NAVSUP 424 Program Management Plan
February 27, 1998	Draft Environmental Checklist and Suggested Items for SECNAVINST 5000-2B
February 27, 1998	Integrated Logistics Support (ILS) Assessments Checklist for Environment, Safety and Health
February 27, 1998	Environment, Safety and Health (ESH) Checklist for Acquisition Statement of Work (SOW) and Plans
May 22, 1998	Current Conference Materials Inventory
June 2, 1998	Draft Exhibit Timeline and Gantt Chart
July 1, 1998	Conference Exhibit Support 1998 Navy Pollution Prevention Conference, June 23–25, 1998
July 1, 1998	Pre-Conference Summary of Required Support Services for the San Antonio Conference in August 1998

July 8, 1998	Minutes of Kick-Off Meeting for Delivery Order 50, May 19, 1998
July 9, 1998	Solid Waste Management Strategic Plan Presentation for the Navy Pollution Prevention Conference, June 1998
July 9, 1998	Conference Report, Overview of Topics Covered during Sessions of the Navy Pollution Prevention Conference, June 1998
July 10, 1998	Delivery Order 50 Summary
July 10, 1998	Deliverable Listing for Delivery Orders 38 and 50
July 15, 1998	Independent Review and Analysis of Procedure for the Demilitarization and Disposal of Naval Vessels
July 21, 1998	Joint Service Pollution Prevention Conference and Exhibition Planning Meeting of July 14, 1998
July 21, 1998	Chief of Naval Operations (CNO) Banner
July 23, 1998	CHRIMP (Consolidated Hazardous Reutilization in Material Programs) Poster/Banner
July 23, 1998	Trip Report, Center for Naval Analyses, Alexandria, Virginia
July 23, 1998	Revised Handouts and Brochures
August 4, 1998	San Antonio Exhibit Forms
August 5, 1998	Informal Technical Note: GAO Report on Defense Management and the Challenges Facing DOD Implementing Defense Reform Initiatives

September 1, 1998	"Quick Reaction Study", Top Ten Navy Environmental Concerns
September 8, 1998	Desk Reference Book of Display Materials Presented at the Joint Service P2 Conference
September 11, 1998	Technical Report, Development of Integrated Logistic Support (ILS) Assessments Reviewers, Checklist for Environment, Safety and Health (ESH)
September 14, 1998	Updated Brochures/Handouts for San Antonio Conference
September 14, 1998	Color Photo of CHRIMP Poster
September 18, 1998	Comments on Draft Test Site NEPA Memo
September 21, 1998	Trip Report, NAVSUP HQ, Mechanicsburg, PA, NRC/NAVICP Data Gathering
September 21, 1998	Trip Report, Conference Support for NAVSUP 424 at the National Recycling Coalition Conference and Exhibition, Albuquerque, New Mexico
September 25, 1998	Conference Support, Third Annual Joint Armed Service Pollution Prevention Conference and Exhibit, August 25–27, 1998
September 25, 1998	Environmental Logistics and Cost Optimization Support of Joint Vision 2010

Appendix II
OTHER WORKS DONE

June 22, 1995	A. F. Meyer & Associates, Inc. Sub-contract Agreement No. 0290-03
September 1995	A Draft Report Assessing the Potential Environmental Impacts of Ship Recycling at Hunter's Point Shipyard
July 16, 1996	Environmental Considerations in the Acquisition Process
April 15–16, 1997	HSMS Site Survey, Naval Intermediate Maintenance Facility, Pearl Harbor, Hawaii
April 17–18, 1997	HSMS Site Survey, Fleet and Industrial Supply Center, Pearl Harbor, Hawaii
April 15–24, 1997	Addendum to HSMS Site Surveys, Pearl Harbor, Hawaii
April 22–24, 1997	HSMS Site Survey, Pearl Harbor Naval Shipyard, Pearl Harbor, Hawaii
April 21 – June 5, 1997	PHNSY Trip Report, Submitted by Jim Bernier
June 1997	Draft – Chapter 2, Identifying Sub-stitutes for Hazardous Materials
June 1997	Draft – Chapter 3, Selecting a Sub-stitute/Using the Pollution Prevention System
June 4, 1997	Draft – Chapter 4, Implementing the Results of the Substitution Process

June 14, 1997	Draft – Appendix A – Examples of Technical Considerations in the Substitution Process
June 14, 1997	Draft – Appendix B – Using the Pollution Prevention System
June 14, 1997	Draft – Appendix C – Examples of Risk and Economic Analyses Worksheets
June 14, 1997	Draft – Appendix D – Glossary of Terms
June 14, 1997	Draft – Appendix E – List of References
June 24, 1997	Excerpts from Pollution Prevention Afloat Reduction of P-D-680 Type II in Planned Maintenance System
July 15, 1997	Appendix A, DON Environmental Program Presentation
July 25, 1997	Adaptation of Military Standard 882D to Pollution Prevention and System Acquisition
July 25, 1997	Executive Summary
July 25, 1997	Phase III, Final Phase of the Substitution Process
July 25, 1997	The Hazardous Material (HM) Substitution Process
August 4–7, 1997	Second Annual Joint Service Pollution Prevention Conference and Exhibition
August 5, 1997	Sustaining the Mission – Sustaining the Planet. Joint Services Pollution Prevention Conference, San Antonio, Texas
November 6, 1997	Contract N00600-95-D-0290, Purchase Order 50-004684, Delivery Order #0038, Monthly Status Report

December 12, 1997	Contract N00600-95-D-0290, Purchase Order #0038, Monthly Status Report
December 24, 1997	Acquisition Category (ACAT) Program Decision PROC
December 24, 1997	Appendix 2 – Acquisition Categories (ACAT)
December 24, 1997	Memorandum – Minimizing Environmental Policy Impacts on Fleet Operations
December 24, 1997	Navy Acquisition Procedures Supplement Change #97-4, Part 5223, Environment, Conservation, Occupational Safety, and Drug-Free Workplace
December 24, 1997	Charter – DON Environmental Acquisition Procurement Working Group
December 24, 1997	Memorandum – Equipment/Systems Requiring the Unplanned Use of Class I Ozone-Depleting Substances (ODS)
December 24, 1997	Memorandum – Affirmative Procurement Program for Items Containing Recovered Materials
December 24, 1997	Executive Order #12969 – Federal Acquisition and Community Right-to-Know
December 24, 1997	Executive Order #12873 – Federal Acquisition Recycling and Waste Prevention
December 24, 1997	Executive Order #12856 – Federal Compliance with Right-to-Know – Pollution Prevention Requirement

December 24, 1997	Executive Order #12843 – Procuring Requirements and Policies for Federal Agencies for Ozone-Depleting Substances
January 20, 1998	Chapter 1 – The Regional Hazardous Materials Management System
January 20, 1998	Chapter 2 – RHMMS System Overview
February 5, 1998	Contract N00600-95-D-0290, Purchase Order 50-004684, Monthly Status Report
February 10, 1998	User Manual for the Regional Inventory Manager, ARMS Around Hazardous Material
February 10, 1998	RHMMS Functional Description
February 10, 1998	Guidelines for Using RHMMS
February 10, 1998	Appendix A – The Hazardous Inventory Control System (HICS)
February 10, 1998	Appendix B – Hazardous Substances Management System (HSMS)
February 10, 1998	Appendix C – RHMMS – HAZMIN User Guide
February 10, 1998	Appendix D – Glossary
February 10, 1998	Appendix E – References
February 27, 1998	Recommended Clarifications to SECNAVINST 5000-2B Regarding Environmental, Health, and Safety Acquisition Requirements
March 1998	Appendix C – Implications of Existing Demilitarization and Disposal Requirements on DOD Ship Scrapping

April 1998	Appendix A – Report of the Interagency Panel on Ship Scrapping
May 19, 1998	Minutes of Kick-Off Meeting for Delivery Order 50
August 10, 1998	Letter – Contract N00600-95-D-0290 – Purchase Order 50-004684 – Monthly Status Report
September 4, 1998	Letter – Contract N00600-95-D-0290 – Purchase Order 50-004684 – Monthly Status Report

Results of the Feasibility Analyses Performed at Indian Head Division, Naval Surface Warfare Center, to Develop Optimum Value Pollution Prevention Alternatives

Appendix III

AN INDEPENDENT ANALYSIS AND RECOMMENDATIONS – IMPLICATIONS FOR THE DOD ON FUTURE US OIL RESERVE SHORTAGES

Note

This presentation is the sole responsibility of the author. In no way does it represent the sponsorship or official position of the DOD, nor those of the military service. The factual information in it is from published documents, to which appropriate attribution is provided.

Abstract

To a large measure, the scope of this paper is more from a national security standpoint, rather than an environmental one, per se. The real issues raised herein are colored by environmental concerns as to economic impacts of oil prices. The real thrust of the discussion is security. Without implementation of such a policy, the worst case scenario described below becomes a real possibility.

If either rising of oil prices or actual overt interruption of supplies occurs, the national security will force very tough choices on the US. Will the nation take military action to preserve supplies of petroleum or change the nation's lifestyle? These are the underlying questions, which this paper intends to pose as a catalyst for in-depth discussions.

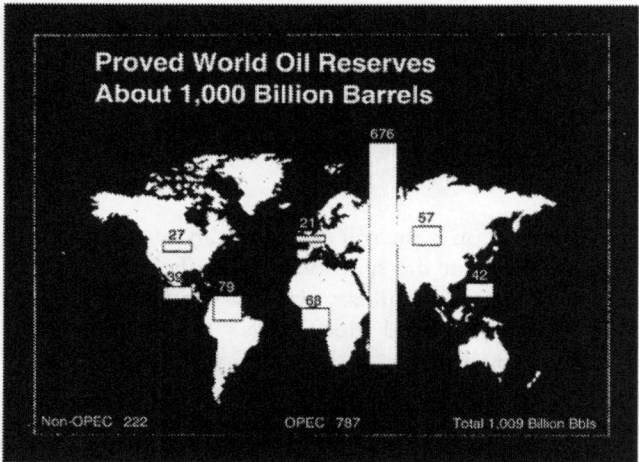

Figure 1: Worldwide Distribution of Oil

Introduction

There have been numerous predictions of impending shortfalls in domestic US oil reserves (*The Economist*, et al). Current estimates of useable oil reserves are 1,009 billion bbls, of which 27 billion bbls (2.7%) are in the US and 676 billion bbls (67.0%) are in the oil-rich nations, which are potentially hostile to the US (*http://www.ecotech.org/pics/sld014.htm*). Figure 1 represents the worldwide distribution of oil reserves (*http://www.ecotech.org/pics/sld014.htm*).

This situation has profound and serious implications to the national security of the US. I, in particular, feel that there are very real possibilities of adverse effects on the ability of the USN and the USAF to project national power as a major element of the military operations other than war (MOOTW). Such operations are increasingly placing major needs for use of air power, including deployment of naval air carriers and USAF air expeditionary forces, as key elements of early MOOTWs. Such efforts are designed to reduce the need for on-site ground forces with attendant possibilities of US casualties.

Scope of Impacts on Forces

There are several possible scenarios associated with the probable lack of domestic oil production. These are discussed in the following sections of this presentation. They range from no impact to worst possible implications. The no impact, or status quo, would include cost increases, resulting in a temporary disruption of acquisition systems. The worst case could result in total inability to logistically support required MOOTW missions.

To better understand the implications of the worst case scenario, the following information is useful regarding the demands for petroleum and petroleum derived items for the air arms.

The normal aviation fuel requirements for the embarked naval air carrier complement is 6,500 bbls per day (GAO/NSIAD-98-1). During MOOTW, such as Desert Storm, the daily need to replenish a single carrier's aviation fuel requirements range from 4,930 to 4,990 bbls per day for the *Theodore Roosevelt* (CVN-71) and USS *America* (CV-66), respectively (GAO/NSIAD-98-1). To meet these needs, the Navy operates conventionally powered tankers and re-supply vessels. These ships also have oil-based fuel requirements. In addition to the nuclear-powered carriers, a carrier task force includes an array of conventionally powered vessels (e.g. AEGIS destroyers, cruisers, etc…) (GAO/NSIAD-98-/). The aircraft carrier itself is not totally dependent on petroleum oil reserves. Even the petroleum-based lubricants are being replaced by synthetic materials. However, production of these synthetics requires energy, much of which is natural gas, or is petroleum-based.

The net effect of the worst case scenario is that the MOOTW air power support capability, other than through long-range missiles, ceases to be effective for both deterrence and actual operations. With reference to the effectiveness of missiles, there are many unanswered questions, both from a military and geo-political aspect.

Implications of Possible Scenarios

In the status quo scenario, the basic assumptions are that the US continues to be the pre-eminent military power through the year 2020. The nation's current dependence on foreign sources of oil increases, but the price thereof remains fairly comparable to 1999 prices. The present consolidation of major sources of oil by US corporation with non-US sources allows for international price and availability stability. Although some sporadic international terrorist acts against the oil powers occur, there is no real effect on the MOOTW or war-fighting capabilities. To some extent, the recent action to increase the US national oil stockpile reserve provides the US forces with a fallback to cover temporary requirements. Energy Secretary Bill Richardson is planning to replace approximately 28 million bbls of oil in the Strategic Petroleum Reserve (SPR) with royalty oil purchased from the

Central Gulf of Mexico.

The results of the worst case scenario were described earlier. These are based on the following assumptions:

1. A united Third World coalition is seeking for geopolitical reasons to enforce the Kyoto Agreement by cutting off oil supplies to those western world nations still contributing excess greenhouse gases, etc. The US is the principal adversary. There are other elements of this antagonistic scenario. These include the Israel/Palestine reserves and Muslim/western world conflicts. (This assumption includes recognition that China is a silent partner of the coalitions stated below.)

2. China, currently or shortly after the embargo, is effective in launching pre-emptive attacks against Taiwan and the Philippines. Appeals for support by these two entities are addressed to the United States. This situation creates a major overriding dilemma for the United States in determining whether or not to exercise the nuclear option. The pros and cons of the answers to that question are beyond the scope of this present discussion.

The Deterrent Option

This section assumes several very real options as preventative measures to avoid the worst case. Among the key assumptions are:

The worst case is recognized and made a component of a comprehensive and combined National Energy and National Security Policy to be implemented by not later than the year 2001. The major elements of implementation of such a policy include:

1. Continuation of the National Strategic Petroleum Reserve, with special procurement of synthetic crude, as detailed below.

2. Reinitiating of a national synthetic fuel program, with emphasis on production of synthetic aviation fuel and diesel fuel. Note:

SASOL Oil (PTY) Ltd., a South Africa-based oil company, is currently producing large quantities of semi-synthetic jet fuel *(http://129.162.25.66/9what/releases/qualify.htm)*. There is not a major R&D need, since the technology for these products is available.

3. The DOE (Department of Energy) has sponsored a number of synthetic fuel demonstration plants. A basis for early restart of such plants or expansion of existing ones gives the United States the capability to initiate full-scale production to meet military needs and ultimately the majority of the existing requirements of the nation.

4. The cost differential of current synthetic fuel compared to domestic oil is a major problem. However, during the time period 2005–2015, it is assumed that the DOD and the DOE will purchase synthetic fuel for petroleum reserve at a price that takes into account real cost of production and a reasonable return on oil investment.

5. As a precursor to implementing the strategy, it is assumed that the development plan includes provisions for field operational testing of South Africa's semi-synthetic aviation and diesel fuels for US naval and USAF aircraft and aboard surface ship engine systems (both diesel and jet engine powered).

Conclusion and Recommendations

It is concluded that a very real threat to national security exists if there is no recognition of the possibilities associated with the future. There are currently available technologies to deter and minimize the threats. The author recommends the following:

- DOD incorporate into the study programs of the National Defense University programs information on this subject.

- The Joint Chiefs of Staff take the lead in incorporating into DOD and the Armed Services War Games, appropriate exercises leading to both interim actions and longer-term solutions within their responsibilities for adequate logistic support.

- The Secretary of Defense take the lead with the DOE, State, and EPA in sponsoring legislation providing for the establishment and implementation of a new National Energy Security Act by 2002.

Addendum

Update: Kosovo Target System

In the recent past, NATO air attacks against Kosovo's oil system and a threat of an oil embargo emphasize the importance of oil to national security interests. In the past, history has shown that oil today is the secret of success in military operations. Airplanes cannot fly, tanks cannot operate, and our transportation system is totally dependent upon oil. Although nuclear power provides some degree of success in generation of electrical power, portable systems such as airplanes, tanks, trains, etc. require petroleum products. During WWII, air superiority and attacks on the Third Reich's oil system helped defeat the Nazi regime. America's great oil reserves are slowly dwindling as the importance of oil is increasing.

The Importance of Synthetic Fuels

During the oil shortage resulting from the embargoes around the 1980s, considerable interest was given to developing a synthetic fuel program in the United States. Relief from the oil embargo and improvements in conservation measures caused the synthetic fuel program to dwindle and die. In fact, oil liquefactions are not new systems. Virtually, every civilized major western city in the early 1900s used coal gas for illumination purposes. Natural gas replaced artificial gas lighting and artificial gas systems fell into decline. The fact remains, however, that considerable improvements in the Lergi System and other coal conversion processes resulted in the late 1920s. Today, only South Africa has a major coal conversion program. Recently, SASOL has produced aviation jet fuel from coal. Thus, no new real research is needed for production processes. However, still remaining are problems associated with environmental safety and health hazards associated with the process and its waste streams. None of these problems is insurmountable and all can be addressed properly, if undertaken now. The time for action is now and not when a crisis occurs in the year 2005 or 2010.

Appendix IV
EXAMPLES OF PROJECTS

February 27, 1980	Intermediate Report, Site Visit Observations and Findings, January 29 – February 20, 1980, SPC Liquid Spill Episode, SRC Pilot Plant, Ft. Lewis, Washington
March 9, 1980	Topical Report, Evaluation of the Mouse Micronucleus Test as Compared with the In-Vivo Cytogenics Test for Mutagenicity of Synthetic Fuel Materials
June 26, 1980	Impact and Compliance, OSHA Carcinogen Policy
July 2, 1980	Report on SARS Backfit Evaluation, Catalytic, Inc., Solvent Refined Coal Pilot Plant, Wilsonville. Alabama
July 2, 1980	Report on SARS Backfit Evaluation, Exxon Donor Solvent Plant, Baytown, Texas
December 1980	Quarterly Report on the Safety Analysis and Review System for the Department of Energy's Fossil Energy Programs, Phase I
January 1981	Interim Status Report, Occupational Safety and Health Training Program at ETCs
January 1981	Review of the ORNL Engineering

	Evaluation of IITRI Project on Design, Engineering, and Evaluation of Refractory Liners for Slagging Gasifiers
February 1981	Evaluation of Alternative Review Plans for the Department of Energy Safety Analysis and Review System
March 1981	Assessment of Documentation Requirements under DOE 5481.1, Safety Analysis and Review System (SARS)
March 1981	Safety Analysis Review Terms of Reference
March 1981	Safety Analysis and Review System (SARS) Assessment Report
March 1981	Analysis and Comparison of Five Contractor Safety and Health Manuals (EG&G, SRC 11, ORNL, Ashland, and MLGW)
March 1981	Final Report, Safety Analysis and Review System, Phase I
March 1981	Audit and Appraisal Plan for Occupational Safety and Health Program, Phase I

Appendix V
EXAMPLES OF EXCERPTS OF PREVIOUS RECOMMENDATIONS

This Appendix presents some excerpts of the types of work the Company has engaged in over the past twenty or thirty years. It includes references to the Air Force Bioenvironmental Engineers, to the work conducted on various projects, and it gives examples of various types of activities the Company has engaged in.

These short examples represent a good selection on file in our Florida office. In addition, a large number of other items in our files are available in our office in Baton Rouge.

A. Pre-1990

1. Department of Defense, Narrative Summaries of Accidents Involving US Nuclear Weapons, 1950–1980.

2. Criteria for Evaluating Gamma Radiation Exposures from Fallout Following Nuclear Detonations, April 1956.

3. Department of Defense Instruction, "Economic Analysis and Program Evaluation for Resource Management", October 18, 1972.

4. Safety Analysis and Review System (SARS), January 31, 1980.

5. "Broken Arrow" Checklist for Bioenvironmental Engineers, October 1983.

6. Military Standard System Safety Program Requirements, March 30, 1984.

7. OPNAVINST Instruction, Navy System Safety Program,

March/April 1987.

8. OPNAV Instruction – Hazardous Material Control and Management, June 20, 1989.

B. 1990–1994

1. Review and Analysis of Draft – DODI 4715.cc.

2. Guidance Manual for Selection/Substitution of Less Hazardous Materials, February 28, 1992.

3. Possible Incentives to Include HMC&M in Acquisition Contracts, May 14, 1992.

4. Information on USAF P2 Policies and Procedures, and Air Combat Command (ACC) Contract, February 8, 1994.

5. Pollution Prevention in Weapon System Acquisition – "A How-to Handbook for Institutionalizing Pollution Prevention in Weapon System Acquisitions", December 30, 1994.

C. 1994–2000

1. DOD Acquisition Deskbook's Reference Set Information Collection Guide.

2. Environmental Planning and Analysis, January 22, 1995.

3. OPNAV Instruction – Pollution Prevention Program (P2P) for Navy Systems, February 24, 1995.

4. Handbook for Integrating Pollution Prevention Into Weapon System Acquisition, February 24, 1995.

5. Weapon System Pollution Prevention Guide for SA/ALC Single Managers, March 1995.

6. Revised Implementing Guidance for Executive Order 12856, "Federal Compliance with Right-to-Know Laws and Pollution Prevention Requirements", April 25, 1995.

7. Naval Supply Systems Command Pollution Prevention

Measures of Success, September 22, 1995.

8. Defense Acquisition Deskbook Informational Briefing, January 8, 1996.

9. Draft – Department of Defense Directive, Defense Acquisition, January 30, 1996.

10. Thoughts on Environment as a Consideration in Naval Warfare Doctrine, May 21, 1996.

11. Environmental Professional Career Training Guide for Civilians, May 1997.

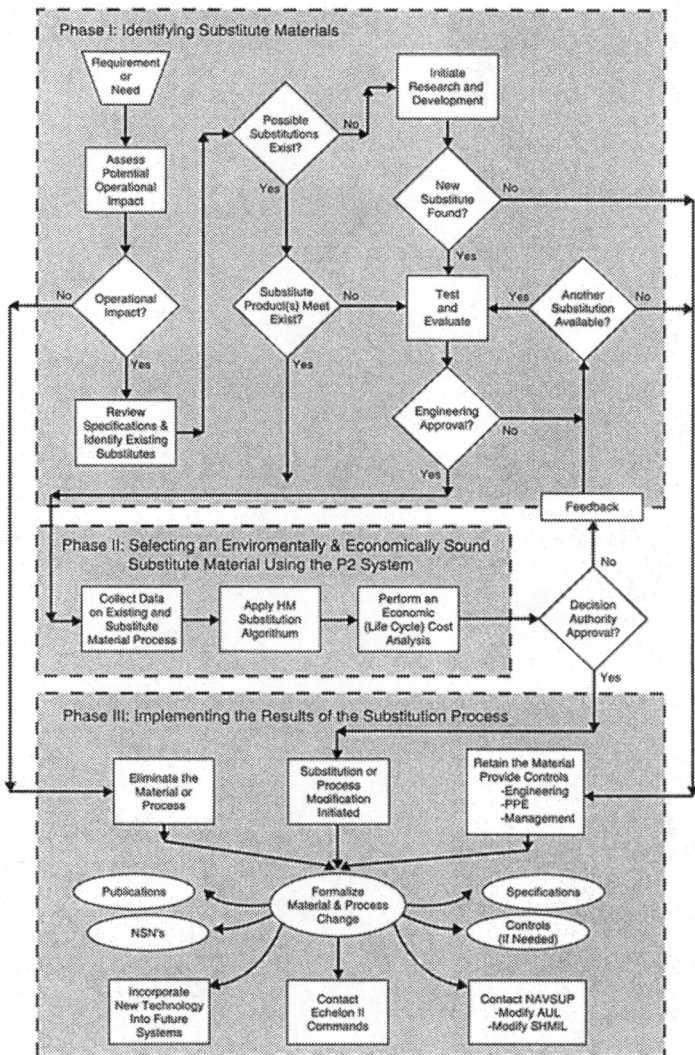

Figure 2: Flow Chart of the Hazardous Material Substitution Process

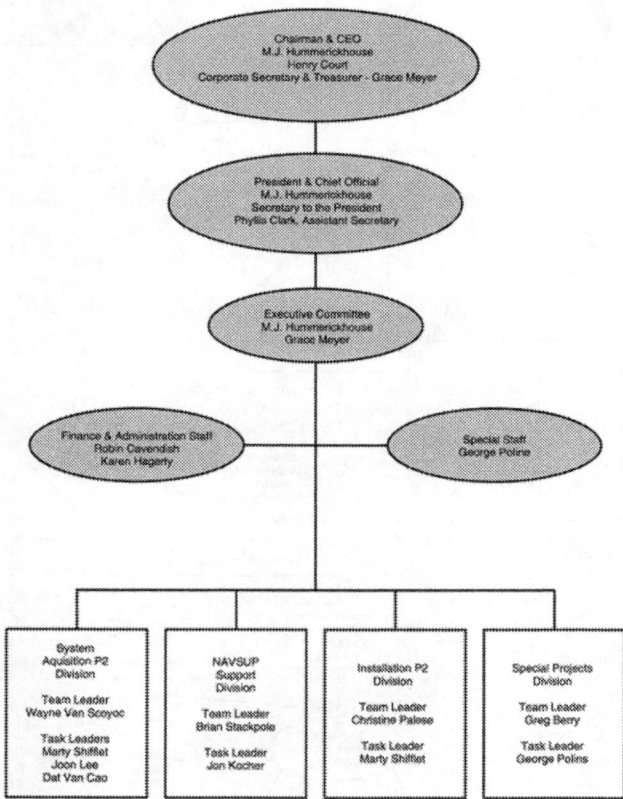

Figure 3: A.F. Meyer & Associates, Inc. Organization Chart